MATHEMATICAL PARADISE

Paul Chika Emekwulu

Mathematical Paradise:
Getting to Know Triangular Numbers, Book Two

Copyright © 2016 Paul Chika Emekwulu

Editing by Christine Rice Publishing Services

Cover Design by EJR Digital Art

Printed in the United States of America

Contents

Preface

Content Organization

This book is divided into eight short chapters.

Chapter 1 exposes us to a summation strategy for first n triangular numbers using the sigma notation.

Chapter 2 explores the sum of first n triangular numbers using the summation notation.

Chapter 3 justifies $\dfrac{n(n+1)(n+2)}{6}$ as the formula for the sum of first n triangular numbers and is inspired by the summation strategies for the following sets of numbers:

first n positive even integers whose first term is equal to 2,

first n whole numbers,

first n positive odd numbers,

first n positive odd integers whose first term is equal to 3,

first n natural numbers,

first n Fibonacci numbers, and first n triangular numbers.

For the later, two cases are considered: when n is odd, and when n is even.

The chapter then unveils the sum of first n triangular numbers.

Chapter 4 discusses the derivation of $\dfrac{n(n+1)(2n+1)}{6}$ from basic or

Paul Emekwulu

first principles.

Chapter 5 discusses another strategy for summing first n triangular numbers. In this investigation, it considers odd and even cases.

Chapter 6 explores properties of triangular numbers. Thirteen such properties are listed. At least four examples are listed for each property.

Chapter 7 contains 14 trial questions on the subject of triangular numbers. It also lists 13 challenge questions.

Chapter 8 contains solutions to independent trial questions.

In chapters 1, 2, 5, and 6, the reader is encouraged to search for counter examples.

Following is the appendix with combination tables of the first 50 triangular numbers, Fibonacci and Lucas numbers. The list of seminars that I present and my internet presence are also featured in this chapter.

The subject of triangular numbers has provided me an opportunity not only to challenge my mind and make a meaningful contribution to the mathematical community, but also to confirm the obvious— the beauty, the elegance, the structure, and the harmony inherent in mathematics.

June 2012
Paul Chika Emekwulu
Norman, Oklahoma
United States of America

Praises for *Mathematical Paradise:*
Getting to Know Triangular Numbers

I found the volume of *Getting to Know Triangular Numbers, Book Two* to be an excellent mathematical study of triangular numbers. All mathematicians—teachers, researchers, and theoreticians—are certain to find it stimulating as the thread of logic weaves its way throughout this topic, bringing together other types of numbers, and creating patterns and relationships that are mathematical beauty to behold.

I do know of books that deal with triangular numbers as a small chapter, unit, or lesson, mostly in a descriptive context, but I am not aware of any that delve into the patterns and relationships as deeply or thoroughly as your book does. I do confess that I have not truly researched the subject either. The mathematics is sound and well-developed, and the objectives listed for each article greatly support the topic of triangular numbers.

Penny Jackson
Lawton Public Schools
Secondary Education/Curriculum Specialist

CHAPTER 1

Summation Formulas for First n Triangular Numbers, Part 1

Objectives

At the end of the lesson, the students should be able to:

- Derive $\sum_{k=1}^{\frac{n+1}{2}}(2k-1)^2$ as the sum of odd numbers of first n triangular numbers.

- Use examples to confirm $\sum_{k=1}^{\frac{n+1}{2}}(2k-1)^2$ as the sum of odd numbers of first n triangular numbers.

- Use $\frac{1}{6}(n)(n+11)(n+2)$ to confirm $\sum_{k=1}^{\frac{n+1}{2}}(2k+1)^2$ as the sum of odd numbers of first n triangular numbers.

- Derive $\sum_{k=1}^{\frac{n}{2}}(2k)^2$ as the sum of even numbers of first n triangular numbers.

- Use examples to confirm $\sum_{k=1}^{\frac{n}{2}}(2k)^2$ as the sum of even numbers of first n triangular numbers.

Introduction

A search for rules for adding series of numbers of the real number system did not start with triangular numbers. If an arithmetic sequence is given in terms of first and last terms, the sum (S_n) is given by:

$$S_n = \frac{n}{2}(a + L).$$

But if it is given in terms of the first term and the difference between the terms of the sequence, either implicitly or explicitly, then

$$S_n = \frac{n}{2}\left[a + a + (n-1)d\right]$$

$$= \frac{n}{2}\left[2a + (n-1)d\right]$$

For first n even numbers,

$$S_n = \frac{n}{2}\left[0 + 0 + 2(n-1)\right]$$

$$= \frac{n}{2}\left[2(n-1)\right]$$

$$= n(n-1)$$

$$= n^2 - n$$

EXAMPLE 1:

Find the sum of 0, 2, 4, 6, and 8.

Paul Emekwulu

$S_5 = 5^2 - 5 = 25 - 5 = 20.$

For first n positive even integers whose first term is equal to 2,

$$S_n = \frac{n}{2}[2 + 2 + 2(n-1)]$$

$$= \frac{n}{2}(4 + 2n - 2)$$

$$= \frac{n}{2}(2n + 2) = n(n+1) = n^2 + n$$

EXAMPLE 2:
Find the sum of 2, 4, 6, 8, and 10.

$S_5 = 5^2 + 5 = 25 + 5 = 30.$

For first n whole numbers,

$$S_n = \frac{n}{2}[a + a + (n-1)d]$$

$$= \frac{n}{2}[0 + 0 + (n-1)1]$$

$$= \frac{n}{2}(n-1)$$

EXAMPLE 3:
Find the sum of 0, 1, 2, 3, and 4.

$$S_5 = \frac{5}{2}(5-1) = 10$$

For first n positive odd integers,

$$S_n = \frac{n}{2}\{1+1+2(n-1)\}$$

$$= \frac{n}{2}(2+2n-2)$$

$$= \frac{n}{2}(2n)$$

$$= n^2.$$

EXAMPLE 4:

Find the sum of 1, 3, 5, 7, and 9.

$S_5 = 5^2 = 25.$

For first n positive odd integers whose first term is equal to 3,

$$S_n = \frac{n}{2}[3+3+2(n-1)]$$

$$= \frac{n}{2}[6+2n-2]$$

$$= \frac{n}{2}(4+2n)$$

$$= n^2 + 2n$$

EXAMPLE 5:

Find the sum of 3, 5, 7, 9, and 11.

$S_5 = 5^2 + 2(5) = 35.$

It has been conventional to use $\dfrac{n(n+1)}{2}$ while adding first n natural

Paul Emekwulu

numbers.

By substituting for the first term in a particular set, we can find a specific rule for summing n terms of that particular set that generally obeys the rule:

$$. \frac{n}{2}(a+L).$$

For the set of natural or counting numbers, $a = 1$, $d = 1$,

$$S_n = \frac{n}{2}\left[a+a+(n-1)d\right]$$

$$= \frac{n}{2}\left[1+1+(n-1)1\right]$$

$$= \frac{n}{2}\left[2+(n-1)\right] = \frac{n}{2}(n+1)$$

EXAMPLE 6:

Find the sum of 1, 2, 3, 4, and 5.

$$S_5 = \frac{5}{2}(5+1) = \frac{5}{2}(6) = 15$$

Each partial sum of first n natural numbers is a triangular number.

Take a look:

$S_1 = 1 = 1$

$S_2 = 1 + 2 = 3$

$S_3 = 1 + 2 + 3 = 6$

$S_4 = 1 + 2 + 3 + 4 = 10$

$S_5 = 1 + 2 + 3 + 4 + 5 = 15$

$S_6 = 1 + 2 + 3 + 4 + 5 + 6 = 21$

$S_7 = 1 + 2 + 3 + 4 + 5 + 6 + 7 = 28$

$S_8 = 1 + 2 + 3 + 4 + 5 + 6 + 7 + 8 = 36$

Sum of First *n* Fibonacci Numbers

The sum of first *n* terms of the Fibonacci sequence can be found as follows:

$S_1 = 1 = 2 - 1 = u_3 - 1$

$S_2 = 1 + 1 = 2 = 3 - 1 = u_4 - 1$

$S_3 = 1 + 1 + 2 = 4 = 5 - 1 = u_5 - 1$

$S_4 = 1 + 1 + 2 + 3 = 7 = 8 - 1 = u_6 - 1$

$S_5 = 1 + 1 + 2 + 3 + 5 = 12 = 13 - 1 = u_7 - 1$

$S_6 = 1 + 1 + 2 + 3 + 5 + 8 = 20 = 21 - 1 = u_8 - 1$

What we are saying here is that, because of the above, the following are true:

$S_1 = u_3 - 1 = u_{(1+2)} - 1$

$S_2 = u_4 - 1 = u_{(2+2)} - 1$

$S_3 = u_5 - 1 = u_{(3+2)} - 1$

$S_4 = u_6 - 1 = u_{(4+2)} - 1$

$S_5 = u_7 - 1 = u_{(5+2)} - 1$

$S_6 = u_8 - 1 = u_{(6+2)} - 1$

Paul Emekwulu

Generally, by inspection,

$S_n = u_{n+2} - 1.$

Again, let u_n represent the nth Fibonacci number.

Then, the following are true:

$S_1 = 1 = 1 = 0 + 1 = 1$	$1 + 0 = S_1 = u_1 + S_0$
$S_2 = 1 + 1 = 2 = 1 + 1 = 2$	$1 + 1 = S_2 = u_2 + S_1$
$S_3 = 1 + 1 + 2 = 4 = 2 + 2 = 4$	$2 + 2 = S_3 = u_3 + S_2$
$S_4 = 1 + 1 + 2 + 3 = 7 = 3 + 4 = 7$	$3 + 4 = S_4 = u_4 + S_3$
$S_5 = 1 + 1 + 2 + 3 + 5 = 12 = 5 + 7 = 12$	$5 + 7 = S_5 = u_5 + S_4$

Table 1: Sum of First n Fibonacci Numbers

$S_1 = u_1 + S_0 = S_1 - S_0 = u_1$
$S_2 = u_2 + S_1 = S_2 - S_1 = u_2$
$S_3 = u_3 + S_2 = S_3 - S_2 = u_3$
$S_4 = u_4 + S_3 = S_4 - S_3 = u_4$
$S_5 = u_5 + S_4 = S_5 - S_4 = u_5$

Generally, $S_n - S_{n-1} = u_n$ implies $S_n = S_{n-1} + u_n.$

So, then $S_n = S_{n-1} + u_n$

But

$$S_{n-1} + u_n = u_{(n-1)+2} - 1 + u_n$$
$$= \left(u_{n+1} - 1 \right) + u_n$$
$$= \left(u_{n+1} + u_n \right) - 1$$
$$= u_{n+2} - 1.$$

Therefore, $S_n = u_{n+2} - 1$.

Sum of First n Triangular Numbers

Now for n triangular numbers, summation can be more complex. In the ensuing chapters, we will investigate strategies we could use in summing up first n triangular numbers. Some of these strategies involve grouping of triangular numbers as sums to form squares. This is primarily based on the fact that the sum of two consecutive triangular numbers is a square number. Because each of the above partial sums is an arithmetic sequence, each of their sums (S_n) of first n natural numbers can be obtained by using:

$$S_n = \frac{n}{2}[a + (n-1)d]$$

where d is the common difference.

This is not so with triangular numbers. Summing up triangular numbers is more complex and provides more options. We shall investigate these in the forthcoming chapters.

Talking of sigma notation in summing up first n triangular numbers involves lower and upper bounds or lower and upper limits. Since we are summing up a specific number of terms, this means we should start somewhere and end somewhere. Usually, the number of first n triangular numbers being summed up could be odd or even. Therefore, we have two cases to consider:

Paul Emekwulu

Case 1: When n is odd.

Case 2: When n is even.

Let's proceed…

Case 1: When n is Odd

When n is odd, a given set of first n triangular numbers can be added as follows:

n	Sum of first n terms
1	$1 = 1^2 = 1$
3	$1 + 3 + 6 = 1^2 + 3^2 = 10$
5	$1 + 3 + 6 + 10 + 15 = 1^2 + 3^2 + 5^2 = 35$
7	$1 + 3 + 6 + 10 + 15 + 21 + 28 = 1^2 + 3^2 + 5^2 + 7^2 = 84$
9	$1 + 3 + 6 + 10 + 15 + 21 + 28 + 36 + 45 + \ldots = 1^2 + 3^2 + 5^2 + 7^2 + 9^2 + \ldots + (2k-1)^2$

Table 2: Partial Sums of First n Triangular Numbers Expressed as Partial Sums of Squares of First n Triangular Numbers.

Since we are finding the sum of a set of triangular numbers, we have to find the following:

(a) An upper bound or an upper limit.

(b) A lower bound or a lower limit.

Why?

Because bearing in mind that the result of

$$1 + (3+6) + (10+15) + (21+28) + \ldots$$

are sums of squares of first n positive odd integers. S_n in turn has to be expressed in terms of squares of positive odd integers (in a general form) and then finally in terms of n.

16

Now, let us find the upper and lower bounds.

Computing the Upper Bound

The last term and, therefore, the largest of the partial sums in Table 2 can be written generally as:

$(2k-1)^2$, $k = \{1, 2, 3, 4, 5, 6...\}$.

This will help us in determining the general form of the upper bound, which has to be expressed in terms of n.

n	n^2	k	$n-k$	$(2k-1)^2$
1	1	1	0	1
3	9	2	1	9
5	25	3	2	25
7	49	4	3	49
9	81	5	4	81
11	121	6	5	121
13	169	7	6	169

Table 3: **Finding the General Form of an Upper Bound**

But $n^2 = (2k-1)^2$.

Taking the square root on both sides, we have:

$n = 2k - 1$ implies $n + 1 = 2k$.

From here, $k = \dfrac{n+1}{2}$.

This becomes the upper bound.

Computing the Lower Bound

The smallest positive odd integer, as well as the first term in each of the partial sums, is 1^2.

Therefore, $(2k-1)^2 = 1$ implies $2k=2$ implies $k=1$.

Therefore, $k = 1$ is the lower bound.

- $(1) = [(2\times1)-1]^2$

- $(1) + (3+6) = [(2\times1)-1]^2 + [(2\times2)-1]^2$

- $(1) + (3+6) + (10+15) = [(2\times1)-1]^2 + [(2\times2)-1]^2 + [(2\times3)-1]^2$

- $(1) + (3+6) + (10 + 15) + (21 + 28) = [(2\times1)-1]^2 + [(2\times2)-1]^2 + [(2\times3)-1]^2 + [(2\times4)-1]^2$

Therefore, $S_n = \sum_{k=1}^{\frac{n+1}{2}} (2k-1)^2$

Consequently, if n is odd, S_n can be found as follows:

$$S_n = \sum_{k=1}^{\frac{n+1}{2}} (2k-1)^2$$

$$= \sum_{k=1}^{\frac{n+1}{2}} \left(4k^2 - 4k + 1\right)$$

$$= 4\left\{ \sum_{k=1}^{\frac{n+1}{2}} k^2 - \sum_{k=1}^{\frac{n+1}{2}} \right\} + \sum_{k=1}^{\frac{n+1}{2}} 1$$

$$= 4\left\{\frac{(k)(k+1)(2k+1)}{6}\right\} - 4\left\{\frac{(k)(k+1)}{6}\right\} + k$$

$$= \left\{\frac{2(k^2+k)(2k+1)}{3}\right\} - \left\{\frac{2k^2+1}{1}\right\} + \frac{k}{1}$$

$$= \frac{4k^3 + 6k^2 + 2k - 3(2k^2+2k)+3k}{3}$$

$$\frac{4k^3 + 6k^2 + 2k - (6k^2+6k)+3k}{3}$$

$$= \frac{4k^3 + 2k - 6k + 3k}{3}$$

$$= \frac{4k^3 - k}{3} = \frac{k(4k^2-1)}{3}$$

$$= \frac{1}{3}(k)(2k-1)(2k+1)$$

By substitution in $\frac{1}{3}(k)(2k-1)(2k+1)$, we have:

Since $k = \frac{n+1}{2}$,

$$\frac{1}{3}(k)(2k-1)(2k+1) = \left(\frac{n+1}{2}\right)\left[2\left(\frac{n+1}{2}\right)-1\right]\left[2\left(\frac{n+1}{2}\right)+1\right]$$

$$= \frac{1}{3}\left(\frac{n+1}{2}\right)(n)(n+2)$$

$$= \frac{1}{6}(n)(n+1)(n+2)$$

Therefore, $S_n = \frac{1}{6}(n)(n+1)(n+2)$.

Verifying Our Result with Examples

EXAMPLE 1:

Consider the numbers 1, 3, 6, 10, and 15.

$$S_n = \sum_{k-1}^{\frac{n+1}{2}}(2k-1)^2 = \frac{1}{3}(k)(2k-1)(2k+1)$$

By substitution, since $n = 5$, $\frac{n+1}{2} = 3$.

$$S_5 = \frac{1}{3}(3)(5)(7) = 35$$

$$S_n = \frac{n(n+1)(n+2)}{6}$$

When $n = 5$, $S_5 = \dfrac{(5)(6)(7)}{6} = 35$.

EXAMPLE 2:

Consider the numbers 1, 3, 6, 10, 15, 21, and 28.

$$S_n = \sum_{k=1}^{\frac{n+1}{2}} (2k-1)^2 = \frac{1}{3}(k)(2k-1)(2k+1)$$

By substitution, since $n = 7$, $\dfrac{n+1}{2} = 4$.

$$S_7 = \frac{1}{3}(4)(7)(9) = 84$$

$$S_n = \frac{n(n+1)(n+2)}{6}.$$

When $n = 7$, $S_7 = \dfrac{(7)(8)(9)}{6} = 84$.

Case 2: When *n* is Even

When n is even, a given set of first n triangular numbers can be added as follows:

n	Sum of first n terms
2	$1 + 3 = 2^2 = 4$
4	$1 + 3 + 6 + 10 = 2^2 + 4^2 = 20$
6	$1 + 3 + 6 + 10 + 15 + 21 = 2^2 + 4^2 + 6^2 = 56$
8	$1 + 3 + 6 + 10 + 15 + 21 + 28 + 36 = 2^2 + 4^2 + 6^2 + 8^2 = 120$
10	$1 + 3 + 6 + 10 + 15 + 21 + 28 + 36 + 45 + 55 + \ldots = 2^2 + 4^2 + 6^2 + 8^2 + 10^2 + \ldots + (2k)^2$

Table 4: Partial Sums of First n Triangular Numbers Expressed as Partial Sums of Squares of First n Odd Numbers.

Similarly, before using the sigma notation to compute S_n, we need an upper bound and a lower bound.

Computing the Upper Bound

The last term and, therefore, the largest positive even integer in each of the partial sums can be written as:

$(2k)^2$, $k = \{1, 2, 3, 4, 5...\}$.

n	n^2	k	$n-k$	$(2k)^2$
2	4	1	1	4
4	16	2	2	16
6	36	3	3	36
8	64	4	4	64
10	100	5	5	100
12	144	6	6	144
16	196	7	7	196
18	324	8	8	324

Table 5: $n^2 = (2k)^2$

The upper bound has to be expressed in terms of n.

Taking the square root on both sides, of $n^2 = (2k)^2$, we have:

$n = 2k$.

From here, $k = \dfrac{n}{2}$.

Therefore, $k = \dfrac{n}{2}$ becomes the upper bound.

    A. $(1+3) = (2 \times 1)^2$

    B. $(1+3) + (6+10) = (2 \times 1)^2 + (2 \times 2)^2$

    C. $(1+3) + (6+10) + (15+21) = (2 \times 1)^2 + (2 \times 2)^2 + (2 \times 3)^2$

D. $(1+3) + (6+10) + (15+21) + (28+36) = (2 \times 1)^2 + (2 \times 2)^2 + (2 \times 3)^2 + (2 \times 4)^2$

Computing the Lower Bound

The smallest positive even integer, as well as the first term in each of the above partial sums (expressed in terms of squares), is 2^2. A positive even integer can be written as $2k$. Its square is then $(2k)^2$.

Therefore, $(2k)^2 = 2^2$.

Taking the square root on both sides, we have: $2k = 2$.

From here, $k = 1$, and that becomes the lower bound.

Therefore, $S_n = \sum_{k=1}^{\frac{n}{2}} (2k)^2 = \sum_{k=1}^{\frac{n}{2}} 4k^2$.

But

$$4 \sum_{k=1}^{\frac{n}{2}} k^2 = 4 \left\{ \frac{(k)(k+1)(2k+1)}{6} \right\}$$

$$= \frac{4\left[\left(\dfrac{n}{2}\right)\left(\dfrac{n}{2}+1\right)\left(2\left(\dfrac{n}{2}\right)+1\right)\right]}{6}, \text{ since } k = \frac{n}{2}.$$

$$= \frac{2}{3}\left(\frac{n}{2}\right)\left(\frac{n+2}{2}\right)(n+1)$$

$$= \frac{n(n+1)(n+2)}{6}.$$

Therefore, $S_n = \frac{1}{6}(n)(n+1)(n+2)$.

Verifying Our Result with Examples

EXAMPLE 1:

Consider the numbers 1, 3, 6, and 10.

$$S_n = \sum_{k=1}^{\frac{n}{2}} (2k)^2$$

By substitution, since $n = 4$, $\frac{n}{2} = 2$

$S_4 = (2 \times 1)^2 + (2 \times 2)^2$

$= 2^2 + 4^2 = 20$

$$S_n = \frac{n(n+1)(n+2)}{6}$$

When $n = 4$, $S_4 = \frac{4(5)(6)}{6} = 20$.

EXAMPLE 2:

Consider the numbers 1, 3, 6, 10, 15, and 21.

24

$$S_n = \sum_{k=1}^{\frac{n}{2}} (2k)^2$$

By substitution, since $n = 6$, $\dfrac{6}{2} = 3$.

$S_6 = (2 \times 1)^2 + (2 \times 2)^2 + (2 \times 3)^2$.

$= 2^2 + 4^2 + 6^2 = 4 + 16 + 36 = 56$

$$S_n = \frac{n(n+1)(n+2)}{6}$$

When $n = 6$, $S_6 = \dfrac{6(7)(8)}{6} = 56$.

CHAPTER 2

Using Sigma Notation to Find the Sum of First n Triangular Numbers, Part 2

Objectives

At the end of the lesson, the students should be able to:

• Use sigma notation to express the sum of first n triangular numbers.

• Use examples to verify the sum of first n triangular numbers.

Introduction

Let us start examining this strategy further by finding various sums of first n triangular numbers, regardless of whether n is even or odd.

n	Sum of first n terms
1	$1 = (1) = 1^2 = 1$
2	$1 + 3 = (1 + 3) = 2^2 = 4$
3	$1 + 3 + 6 = (1) + (3 + 6) = 1^2 + 3^2 = 10$
4	$1 + 3 + 6 + 10 = (1 + 3) + (6 + 10) = 2^2 + 4^2 = 20$
5	$1 + 3 + 6 + 10 + 15 = (1) + (3 + 6) + (10 + 15) = 1^2 + 3^2 + 5^2 = 35$

Table 5: Partial Sums of First n Triangular Numbers.

The above can be separated and classified among:

(a) Even cases.

(b) Odd Cases.

Without this classification, the situation could be blurred and we will not be able to clearly see the logic of the situation.

Even Cases

EXAMPLE 1:

$1 + 3 + 6 + 10 + 15 + 21$

$= (1+3) + (6+10) + (15+21)$

$= 4 + 16 + 36$

$= [6-2(2)]^2 + [6-2(1)]^2 + [6-2(0)]^2$

EXAMPLE 2:

$1 + 3 + 6 + 10 + 15 + 21 + 28 + 36$

$= (1+3) + (6+10) + (15+21) + (28+36)$

$= 4+16+36+64$

$= [8-2(3)]^2 + [8-2(2)]^2 + [8-2(1)]^2 + [8-2(0)]^2$

EXAMPLE 3:

$1 + 3 + 6 + 10 + 15 + 21 + 28 + 36 + 45 + 55$

$= (1+3) + (6+10) + (15+21) + (28+36) + (45+55)$

$= 4+16+36+64+100$

$= [10-2(4)]^2 + [10-2(3)]^2 + [10-2(2)]^2 + [10-2(1)]^2 + [10-(0)]^2$

EXAMPLE 4:

$1+3+6+10+15+21 + 28 + 36 + 45 + 55 + 66 + 78$

$= (1+3) + (6+10) + (15+21) + (28+36) + (45+55) + (66+78)$

27

Paul Emekwulu

$= 4+16+36+64+100+144$

$= [12–2(5)]^2 + [12–2(4)]^2 + [12–2(3)]^2 + [12–2(2)]^2 + [12–2(1)]^2 + [12–2(0)]^2$

The individual terms in the final expressions in examples 1, 2, 3, and 4 above can be generally written as $(n–2k)^2$.

Finding the Lower Bound for Even Cases

Let us consider the sum of n even numbers of first n triangular numbers.

n	Sum of first n terms
2	$1 + 3 = 2^2 = 4$
4	$1 + 3 + 6 + 10 = 2^2 + 4^2 = 20$
6	$1 + 3 + 6 + 10 + 15 + 21 = 2^2 + 4^2 + 6^2 = 56$
8	$1 + 3 + 6 + 10 + 15 + 21 + 28 + 36 = 2^2 + 4^2 + 6^2 + 8^2 = 120$

Table 6: **Sums of Even Numbers of First n Triangular Numbers**

For such a set of even numbers of first n triangular numbers,

$S_n = 2^2 + 4^2 + 6^2 + ... + (n–2k)^2.$

For such a sum of even numbers of first n triangular numbers, the first term is 2^2.

Therefore, for a lower bound, the following condition must exist:.

$(n – 2k)^2 = 2^2$

For the above equation to be true, the following must be true:

$n – 2 = 2k$

From here, $n - 2 = 2k$ implies $k = \dfrac{n-2}{2}$.

n	$n-2$	$\dfrac{n-2}{2}$
2	0	0
4	2	1
6	4	2
8	6	3
10	8	4
12	10	5
14	12	6

Table 7: Finding the General Form of the Lower Bound

Definition: A positive even integer can be written as $2k + 2$, where $k = \{0, 1, 2, 3, 4...\}$.

n	Sum of first n terms
2	$1 + 3 = 2^2 = 4$
4	$1 + 3 + 6 + 10 = 2^2 + 4^2 = 20$
6	$1 + 3 + 6 + 10 + 15 + 21 = 2^2 + 4^2 + 6^2 = 56$
8	$1 + 3 + 6 + 10 + 15 + 21 + 28 + 36 = 2^2 + 4^2 + 6^2 + 8^2 = 120$

Table 8: Sums of Partial Sums of Even Number of First n Triangular Numbers

Partial Sums of Squares of First n Positive Even Integers

The smallest positive even integer, as well as the first term in each of the partial sums in Table 8, is 2^2. Is it not? This is also clearly shown below:

Consider the following and its partial sums.

29

$2^2 + 4^2 + 6^2 + 8^2 + 10^2 + \ldots + (2k\text{-}2)^2$

$2^2 + \ldots + (2k-2)^2$

$2^2 + 4^2 + \ldots + (2k-2)^2$

$2^2 + 4^2 + 6^2 + \ldots + (2k-2)^2$

$2^2 + 4^2 + 6^2 + 8^2 \ldots + (2k-2)^2$

$2^2 + 4^2 + 6^2 + 8^2 + 10^2 \ldots + (2k-2)^2$

Therefore, $(2k + 2)^2 = 2^2$.

By taking the square root on both sides, we have:

$2k + 2 = 2$ implies $2k = 0$, which implies $k = 0$.

Therefore, $k = 0$ is the lower bound.

n	n^2	k	$2k+2$	$(2k+2)^2$
2	4	0	2	4
4	16	1	4	16
6	36	2	6	36
8	64	3	8	64
10	100	4	10	100
10	144	5	12	144
14	196	6	14	196
16	256	7	16	256

Table 9: Finding the General Form of the Upper Bound

The square of the last term in each of the partial sums is equal to n^2.

Therefore, $n^2 = (2k+2)^2$.

$n = 2k + 2$ implies $n - 2 = 2k$ implies $k = \dfrac{n-2}{2}$.

Therefore, $k=\dfrac{n-2}{2}$ becomes the upper bound.

The last term can still be written as $2k+2$.

$\dfrac{2-2}{2} = 0$ implies upper bound, when $n=2$.

$\dfrac{4-2}{2} = 0$ implies upper bound, when $n=4$.

$\dfrac{6-2}{2} = 0$ implies upper bound, when $n=6$.

$\dfrac{8-2}{2} = 0$ implies upper bound when $n=8$.

$\dfrac{10-2}{2} = 0$ implies upper bound, when $n=10$.

Having done this, a pattern is evident. A conclusion we drew earlier has also been confirmed here that for even cases, the upper bound is:

$$\dfrac{n-2}{2}$$

Therefore, for even cases,

$$S_n = (n-0)^2 + (n-2)^2 + (n-4)^2 + \dots + (n-2k)^2$$

and can be expressed with the summation notation as follows:

Paul Emekwulu

$$S_n = \sum_{k=0}^{\frac{n-2}{2}} (n-2k)^2$$

where $k = 0$ is the lower bound and $k = \dfrac{n-2}{2}$ is the upper bound.

EXAMPLE 1

Consider $1 + 3 + 6$.

When $n = 3$, $\dfrac{n+1}{2} = \dfrac{3+1}{2} = 2$.

$$\sum_{k=0}^{\frac{n+1}{2}} (n-2k)^2 = [3 - (2 \times 0)]^2 + [3 - (2 \times 1)]^2$$

$$= 3^2 + 1^2 = 9 + 1 = 10.$$

EXAMPLE 2

Consider $1 + 3 + 6 + 10 + 15$.

When $n = 5$, $\dfrac{n+1}{2} = \dfrac{5+1}{2} = 3$.

$$\sum_{k=0}^{\frac{n+1}{2}} (n-2k)^2 = [3 - (2 \times 0)]^2 + [5 - (2 \times 1)]^2 + [5 - (2 \times 2)]^2$$

$$= 5^2 + 3^2 + 1^2 = 25 + 9 + 1 = 35.$$

EXAMPLE 3:

Consider $1 + 3$.

When $n = 2$, $\dfrac{n-2}{2} = \dfrac{2-2}{2} = \dfrac{0}{2} = 0.$

$S_2 = \displaystyle\sum_{k=0}^{0} (n-2k)^2$

$= [2-(2\times 0)]^2$

$= 2^2 = 4$

EXAMPLE 4:

Consider $1 + 3 + 6 + 10$.

When $n = 4$, $\dfrac{n-2}{2} = \dfrac{4-2}{2} = \dfrac{2}{2} = 1.$

$S_4 = \displaystyle\sum_{k=0}^{1} (n-2k)^2$

$= [4-(2\times 0)]^2 + [4-(2\times 1)]^2$

$= 4^2 + 2^2 = 16 + 4 = 20.$

EXAMPLE 5:

Consider $1 + 3 + 6 + 10 + 15 + 21$.

When $n = 6$, $\dfrac{n-2}{2} = \dfrac{6-2}{2} = \dfrac{4}{2} = 2.$

$S_6 = \displaystyle\sum_{k=0}^{2} (n-2k)^2$

Paul Emekwulu

$$= \left[6-(2\times0)\right]^2 + \left[6-(2\times1)\right]^2 + \left[6-(2\times2)\right]^2$$

$= 6^2+4^2+2^2 = 36 + 16 + 4 = 56.$
EXAMPLE 6:

Consider $1 + 3 + 6 + 10 + 15 + 21 + 28 + 36.$

When $n = 8$, $\dfrac{n-2}{2} = \dfrac{8-2}{2} = \dfrac{6}{2} = 3.$

$$S_8 = \sum_{k=0}^{3} (n-2k)^2$$

$$= [8-(2\times0)]^2 + [8-(2\times1)]^2 + [8-(2\times2)]^2 + [8-(2\times3)]^2$$

$$= 8^2 + 6^2 + 4^2 + 2^2$$

$$= 64 + 36 + 16 + 4 = 120.$$

EXAMPLE 7:

Consider $1 + 3 + 6 + 10 + 15 + 21 + 28 + 36 + 45 + 55.$

When $n = 10$, $\dfrac{n-2}{2} = \dfrac{10-2}{2} = \dfrac{8}{2} = 4.$

$$S_{10} = \sum_{k=0}^{4} (n-2k)^2$$

$$= [10-(2\times0)]^2 + [10-(2\times1)]^2 + [10-(2\times2)]^2 + [10-(2\times3)]^2 + [10-(2\times4)]^2$$

$$= 10^2 + 8^2 + 6^2 + 4^2 + 2^2$$

$$= 100 + 64 + 36 + 16 + 4 = 216.$$

EXAMPLE 8:

Consider $1 + 3 + 6 + 10 + 15 + 21 + 28 + 36 + 45 + 55 + 66 + 78$.

When $n = 12$, $\dfrac{n-2}{2} = \dfrac{12-2}{2} = \dfrac{10}{2} = 5$.

$$S_{12} = \sum_{k=0}^{5}(n-2k)^2$$

$= [12-(2\times0)]^2 + [12-(2\times1)]^2 + [12-(2\times2)]^2 + [12-(2\times3)]^2 + [12-(2\times4)]^2 + [12-(2\times5)]^2$

$= 12^2 + 10^2 + 8^2 + 6^2 + 4^2 + 2^2$

$= 144 + 100 + 64 + 36 + 16 + 4 = 364$.

A Search for Counter Examples

Can you identify any even number of first n triangular numbers

whose sum is not given by $\displaystyle\sum_{k=0}^{\frac{n-2}{2}}(n-2k)^2$?

Finding the Lower Bound for Odd Cases

Let us consider the sum of n odd numbers of first n triangular numbers.

n	Sum of first n terms
1	$1 = 1^2 = 1$
3	$1 + 3 + 6 = 1^2 + 3^2 = 10$
5	$1 + 3 + 6 + 10 + 15 = 1^2 + 3^2 + 5^2 = 35$
7	$1 + 3 + 6 + 10 + 15 + 21 + 28 = 1^2 + 3^2 + 5^2 + 7^2 = 84$
9	$1 + 3 + 6 + 10 + 15 + 21 + 28 + 36 + 45 = 1^2 + 3^2 + 5^2 + 7^2 + 9^2 = 165$

Table 10: Sum of Odd Number of First n Triangular Numbers

Definition: A positive odd integer can be expressed as $2k + 1$,

Paul Emekwulu

where $k = \{0, 1, 2, 3...\}$.

The smallest positive odd integer, as well as the first term in each of the partial sums (sums of squares of positive odd integers) in Table 10, is 1^2. This is clearly shown below:

$1 + 3 + 5 + 7 + 9 + 11 + 13 + 15...$

$1 + ... + (2k - 1)$

$1 + 3 + ... + (2k - 1)$

$1 + 3 + 5 + ... + (2k - 1)$

$1 + 3 + 5 + 7 + ... + (2k - 1)$

$1 + 3 + 5 + 7 + 9 + ... + (2k - 1)$

Consider the following and the partial sums.

$1^2 + 3^2 + 5^2 + 7^2 + 9^2 + ... + (2k-1)^2$

$1^2 + ... + (2k - 1)^2$

$1^2 + 3^2 + ... + (2k - 1)^2$

$1^2 + 3^2 + 5^2 + ... + (2k - 1)^2$

$1^2 + 3^2 + 5^2 + 7^2... + (2k - 1)^2$

$1^2 + 3^2 + 5^2 + 7^2 + 9^2... 2k - 1)^2$

Partial Sums of Squares of First *n* Positive Odd Integers

Therefore, $(2k+1)^2 = 1^2$.

Taking the square root on both sides, $2k + 1 = 1$ implies $k = 0$.

Therefore, $k = 0$ becomes the lower bound.

n	n^2	k	$2k+1$	$(2k+1)^2$
1	1	0	1	1
3	9	1	3	9
5	25	2	5	25
7	49	3	7	49
9	81	4	9	81
11	121	5	11	121
13	169	6	15	169
15	225	7	17	225

Table 11: Finding the Lower Bound

In a general sense, the last term in each of the partial sums is equal to n^2.

Therefore, $n^2 = (2k+1)^2$ (see Table 12).

Taking the square root on both sides, we have:

$n = 2k + 1$ implies $n - 1 = 2k$, which in turn implies that:

$$k = \frac{n-1}{2}$$

$$\frac{1-1}{2} = 0 \text{ implies upper bound when } n = 1.$$

$$\frac{3-1}{2} = 2 \text{ implies upper bound when } n = 3$$

$$\frac{5-1}{2} = 1 \text{ implies upper bound when } n = 5.$$

$$\frac{7-1}{2} = 3 \text{ implies upper bound when } n = 7.$$

Paul Emekwulu

$$\frac{9-1}{2} = 4 \text{ implies upper bound when } n = 9.$$

Therefore, $k = \dfrac{n-1}{2}$ becomes the upper bound.

The above division facts can be generally represented as:

$$k = \frac{n-1}{2}$$

Then, the upper bound for odd cases is:

$$k = \frac{n-1}{2}$$

$$S_n = (n{-}0)^2 + (n{-}2)^2 + (n{-}4)^2 + \ldots + (n{-}2k)^2$$

$$= \sum_{k=0}^{\frac{n-1}{2}} (n - 2k)^2$$

where $k = 0$ is the lower bound and $k = \dfrac{n-1}{2}$ is the upper bound.

EXAMPLE 1:

Consider $1 + 3 + 6$.

When $n = 3$, $\dfrac{n-1}{2} = \dfrac{3-1}{2} = \dfrac{2}{2} = 1$.

$$S_3 = \sum_{k=0}^{1} (n - 2k)^2$$

$$= [3 - (2{\times}0)]^2 + [3 - (2{\times}1)]^2$$

$= 3^2 + 1^2 = 9 + 1 = 10.$

EXAMPLE 2:

Consider $1 + 3 + 6 + 10 + 15.$

When $n = 5$, $\dfrac{n-1}{2} = \dfrac{5-1}{2} = \dfrac{4}{2} = 2.$

$$S_5 = \sum_{k=0}^{2}(n-2k)^2$$

$$= [5 - (2\times0)]^2 + [5 - (2\times1)]^2 + [5 - (2\times2)]^2$$

$$= 5^2 + 3^2 + 1^2 = 25 + 9 + 1 = 35.$$

EXAMPLE 3:

Consider $1 + 3 + 6 + 10 + 15 + 21 + 28.$

When $n = 7$, $\dfrac{n-1}{2} = \dfrac{7-1}{2} = \dfrac{6}{2} = 3.$

$$S_7 = \sum_{k=0}^{3}(n-2k)^2$$

$$= [7 - (2\times0)]^2 + [7 - (2\times1)]^2 + [7 - (2\times2)]^2 + [7 - (2\times3)]^2$$

$$= 7^2 + 5^2 + 3^2 + 1^2 = 49 + 25 + 9 + 1 = 84.$$

EXAMPLE 4:

Consider $1 + 3 + 6 + 10 + 15 + 21 + 28 + 36 + 45.$

When $n = 9$, $\dfrac{n-1}{2} = \dfrac{9-1}{2} = \dfrac{8}{2} = 4.$

$$S_9 = \sum_{k=0}^{4} (n-2k)^2$$

$$= [9-(2\times0)]^2 + [9-(2\times1)]^2 + [9-(2\times2)]^2 + [9-(2\times3)]^2 + [9-(2\times4)]^2$$

$$= 9^2 + 7^2 + 5^2 + 3^2 + 1^2 = 81 + 49 + 25 + 9 + 1 = 165.$$

EXAMPLE 5:

Consider $1 + 3 + 6 + 10 + 15 + 21 + 28 + 36 + 45 + 55 + 66.$

When $n = 11$, $\dfrac{n-1}{2} = \dfrac{11-1}{2} = \dfrac{10}{2} = 5.$

$$S_{11} = \sum_{k=0}^{5} (n-2k)^2$$

$$= [11-(2\times0)]^2 + [11-(2\times1)]^2 + [11-(2\times2)]^2 + [11-(2\times3)]^2 + [11-(2\times4)]^2 + [11-(2\times5)]^2$$

$$= 11^2 + 9^2 + 7^2 + 5^2 + 3^2 + 1^2 = 121 + 81 + 49 + 25 + 9 + 1 = 286.$$

A Search for Counter Examples

1. Can you identify any even number of first n triangular numbers whose sum is not equal to:

$$\sum_{k=0}^{\frac{n-2}{2}} (n-2k)^2 \ ?$$

2. Can you identify any odd number of first n triangular numbers whose sum is not equal to: $\displaystyle\sum_{k=0}^{\frac{n-1}{2}} (n-2k)^2$

CHAPTER 3

Justifying $\dfrac{(n)(n+1)(n+2)}{6}$ as the Formula for the Sum of First n Triangular Numbers

Objectives

At the end of the lesson, the students should be able to:

•Justify the formula for the sum of first n triangular numbers.

Introduction

Let us start by considering some partial sums of first n triangular numbers.

Doing so, we have:

1

$1 + 3$

$1 + 3 + 6$

$1 + 3 + 6 + 10$

$1 + 3 + 6 + 10 + 15$

$1 + 3 + 6 + 10 + 15 + 21$

We can use $\dfrac{(n)(n+1)(n+2)}{6}$ to find the sums of n terms of each of the above subsets.

Let $\dfrac{(n)(n+1)(n+2)}{6} = S_n$.

Paul Emekwulu

Doing so, we have:

$1 = 1 = S_1$

$1 + 3 = 4 = S_2$

$1 + 3 + 6 = S_3$

$1 + 3 + 6 + 10 = S_4$

$1 + 3 + 6 + 10 + 15 = S_5$

$1 + 3 + 6 + 10 + 15 + 21 = S_6$

$1 + 3 + 6 + 10 + 15 + 21 + 28 = S_7$

$1 + 3 + 6 + 10 + 15 + 21 + 28 + 36 = S_8$

As a part of the investigation, let us find the following for some values of n:

(a) S_n

(b) S_{n-1}

(c) $S_{n-1} + S_n$

Take a look:

$5 - 4 = 1$

$14 - 10 = 4$

$30 - 20 = 10$

$55 - 35 = 20$

In other words, the above are equal to:

$(1+4) - 4 = 1$

$(4+10) - 10 = 4$

$(10+20) - 20 = 10$

$(20+35) - 35 = 20$

Generally, we can represent the above as follows:

$(S_1 + S_2) - S_2 = S_1$

$(S_2 + S_3) - S_3 = S_2$

$(S_3 + S_4) - S_4 = S_3$

$(S_4 + S_5) - S_5 = S_4$

Generally, $(S_{n-1} + S_n) - S_n = S_{n-1}$.

Let us extract the subscripts and then display them as shown in

$n-1$	n	k	$n + (n-1) - n = k$
1	2	1	$2+ (2-1) - 2 = 1$
2	3	2	$3+ (3-1) - 3 = 2$
3	4	3	$4+ (4-1) - 4 = 3$
4	5	4	$5+ (5-1) - 5 = 4$
5	6	5	$6+ (6-1) - 6 = 5$
6	7	6	$7+ (7-1) - 7 = 6$

Table 12.

$n-1$	n	k	$n + (n-1) - n = k$
1	2	1	$2+ (2-1) - 2 = 1$
2	3	2	$3+ (3-1) - 3 = 2$
3	4	3	$4+ (4-1) - 4 = 3$
4	5	4	$5+ (5-1) - 5 = 4$
5	6	5	$6+ (6-1) - 6 = 5$
6	7	6	$7+ (7-1) - 7 = 6$

Paul Emekwulu

Table 12: $n + (n-1) - n = k$

$n-1$	n	k	$n + (n-1) - n = k$
1	2	1	$2 + (2-1) - 2 = 1$
2	3	2	$3 + (3-1) - 3 = 2$
3	4	3	$4 + (4-1) - 4 = 3$
4	5	4	$5 + (5-1) - 5 = 4$
5	6	5	$6 + (6-1) - 6 = 5$
6	7	6	$7 + (7-1) - 7 = 6$

From

Table 12, we notice the following relationship:

$n + (n-1) - n = k$ (i.e. $k = n - 1$).

Can we derive S_n from S_{n-1}?

Actually, $(S_{n-1} + S_n) - S_n = S_{n-1}$.

The answer is, therefore, yes, we can derive S_n from S_{n-1}.

This shapes the ground for continued investigation, and a very important part of this investigation is simplifying $S_{n-1} + S_n$.

In addition, the following should be noted:

a) $(S_n + S_{n-1}) - S_n = \dfrac{n(n+1)(2n+1)}{6}$

b) $S_n = \dfrac{n(n+1)(2n+1)}{6}$.

c) $(S_n + S_{n-1}) - S_n = S_n$

Then, by substitution, we have:

$$(S_n + S_{n-1}) - S_n = \frac{n(n+1)(2n+1)}{6} - \frac{n(n+1)(2n+1)}{6}$$

$$= \frac{(n^2+n)(2n+1)}{6} - \frac{n(n+1)(n+2)}{6}$$

$$= \frac{(n^2+n)(2n+1) - [(n^2+n)(n+2)]}{6}$$

$$= \frac{2n^3 + n^2 + 2n^2 + n - n^3 - 2n^2 - n^2 - 2n}{6}$$

$$= \frac{2n^3 - n^3 + 3n^2 - 3n^2 + n - 2n}{6}$$

$$= \frac{n^3 - n}{6} = \frac{n(n^2-1)}{6}$$

$$= \frac{(n)(n-1)(n+1)}{6}, \ n \geq 2.$$

By mathematical induction,

If $S_n = \dfrac{n\,(n-1)(n+1)}{6}, \ n \geq 2$

is true for n, it is true for $n+1$.

Therefore, $S_{n+1} = \dfrac{(n+1)(n+1-1)(n+1+1)}{6}.$

$$= \frac{(n+1)(n)(n+2)}{6}$$

Paul Emekwulu

$$= \frac{(n)(n+1)(n+2)}{6}, \; n \geq 1$$

Validating Our Result

Let us substitute some values for n to validate our result.

We will eventually explore a relationship between:

$$\frac{(n)(n-1)(n+1)}{6} \quad \text{and} \quad \frac{(n)(n+1)(n+2)}{6}$$

When $n = 1$, $\dfrac{(n)(n-1)(n+1)}{6} = 0$ and $\dfrac{(n)(n+1)(n+2)}{6} = 1$.	
When $n = 2$, $\dfrac{(n)(n-1)(n+1)}{6} = 1$ and $\dfrac{(n)(n+1)(n+2)}{6} = 4$.	
When $n = 3$, $\dfrac{(n)(n-1)(n+1)}{6} = 4$ and $\dfrac{(n)(n+1)(n+2)}{6} = 10$.	
When $n = 4$, $\dfrac{(n)(n-1)(n+1)}{6} = 10$ and $\dfrac{(n)(n+1)(n+2)}{6} = 20$.	
When $n = 5$, $\dfrac{(n)(n-1)(n+1)}{6} = 20$ and $\dfrac{(n)(n+1)(n+2)}{6} = 35$.	

Table 13: **Validating Our Result**

Look at the same information in a different format.

46

n	$\dfrac{(n)(n-1)(n+1)}{6}$	$\dfrac{(n)(n+1)(n+2)}{6}$
1	0	1
2	1	4
3	4	10
4	10	20
5	20	35

Table 14: Still on Validating Result

As a result of the above substitution in Table 14, two sets of numbers, A and B, were generated.

Take a look:

$A = \{0, 1, 4, 10, 20, 35...\}$

$B = (1, 4, 10, 20, 35...\}$

But $\{0, 1, 4, 10, 20, 35...\}$ is not consistent with $\{1, 4, 10, 20,$

$35...\}$ as sums of first n triangular numbers.

From a previous chapter, $\dfrac{(n)(n-1)(n+1)}{6}$ is already a familiar

expression, but while substituting various values of n in

$\dfrac{(n)(n-1)(n+1)}{6}$

the set $\{0, 1, 4, 10...\}$ emerged, and this set contains the subset $\{1,$

47

Paul Emekwulu

4, 10, 20...},

which is the set of partial sums of first n triangular numbers.

Now let $\dfrac{(n)(n-1)(n+1)}{6} = p$ ....(i)

$\dfrac{(n)(n+1)(n+2)}{6} = q$ ....(ii)

Our thinking here is that we can derive Equation (ii) from Equation (i) using mathematical induction. Our objective is, therefore, deriving

$$\dfrac{(n)(n+1)(n+2)}{6} \text{ from } \dfrac{(n)(n-1)(n+1)}{6}.$$

For the same values of n,

$$\dfrac{(n)(n-1)(n+1)}{6} \text{ and } \dfrac{(n)(n+1)(n+2)}{6}$$

are yielding different numerical results

(See Table 13 and Table 14).

Yes, of course, that is expected.

But for different values of n,

$$\dfrac{(n)(n-1)(n+1)}{6} \text{ and } \dfrac{(n)(n+1)(n+2)}{6}$$

start to yield the same results.

Because of this, let us represent the n in $\dfrac{(n)(n-1)(n+1)}{6}$ by n'

while the n in

$$\frac{(n)(n+1)(n+2)}{6}$$

remains unchanged.

the n in $\dfrac{(n)(n-1)(n+1)}{6}$ by n'.

The quantities $\dfrac{(n)(n+1)(n+2)}{6}$ and $\dfrac{(n')(n'-1)(n'+1)}{6}$

start to yield the same results when $n' = 2$ and $n = 1$
(See Table 15).

n	$\dfrac{(n)(n+1)(n+2)}{6}$	n'	$\dfrac{n'(n'-1)(n'+1)}{6}$
1	1	2	1
2	4	3	4
3	10	4	10
4	20	5	20
5	35	6	35
6	56	7	56
7	84	8	84

Table 15: The Same Numerical Results for Different Values of n

n	n'	$n'-n$
1	2	1
2	3	1
3	4	1
4	5	1
5	6	1

Table 16: The Relationship between n and n'

From Table 16, $n' - n = 1$ implies that $n' = n + 1$.

By substituting for n' in $\dfrac{(n')(n'-1)(n'+1)}{6}$, we have:

$$\frac{(n')(n'-1)(n'+1)}{6} = \frac{(n+1)[(n+1)-1][(n+1)+1]}{6}$$

$$= \frac{n(n+1)(n+2)}{6}$$

$$(S_n + S_{n+1}) - S_n = \left[\frac{n(n+1)(2n+1)}{6}\right] - \left[\frac{n(n+1)(n+2)}{6}\right]$$

$$\frac{(n^2+n)(2n+1) - (n^2+n)(n+2)}{6}$$

$$= \frac{2n^3 + n^2 + 2n^2 + n - n^3 - 2n^2 - n^2 - 2n}{6}$$

$$= \frac{\left(2n^3 - n^3\right) + \left(n^2 - n^2\right) + \left(2n^2 - 2n^2\right) + \left(n - 2n\right)}{6}$$

$$= \frac{n^3 - n}{6} = \frac{n\left(n^2 - 1\right)}{6} = \frac{(n-1)(n)(n+1)}{6}.$$

$$= \frac{(n)(n-1)(n+1)}{6}$$

A Search for Counter Examples

Can you identify any subset of first n triangular numbers whose sum cannot be found by using the equation below?

$$S_n = \frac{(n)(n+1)(n+2)}{6}$$

CHAPTER 4

Deriving a Formula for the Sum of First *n* Triangular Numbers from Basic Principles

Objectives

At the end of the lesson, the students should be able to:

• Use basic principles to derive

$$\frac{n(n+1)(2n+1)}{6}$$

as partial sums of first *n* triangular numbers.

Introduction

In this chapter, our goal is to derive:

$$\frac{n(n+1)(2n+1)}{6}$$

from basic principles.

This is a formula for finding the partial sums of first *n* triangular numbers.

Look at the following partial sums:

$1 = 1$

$1 + 3 = 4$

$1 + 3 + 6 = 10$

$1 + 3 + 6 + 10 = 20$

$1 + 3 + 6 + 10 + 15 = 35$

$1 + 3 + 6 + 10 + 15 + 21 = 56$

Let's find the partial sums of the above sums.

Doing so, we have:

$1 = 1$ $10 + 20 = 30$

$1 + 4 = 5$ $20 + 35 = 55$

$4 + 10 = 14$ $35 + 56 = 91$

Multiplying Partial Sums by 6

Now, multiplying each of the above sums by 6 transforms each of them into a number q. Each q exists as a number mn, where $n \geq m$. Each mn can further be expressed in the form, $a \times b \times c$.

Therefore, q or $mn = a \times b \times c = (a \times b)(a+b)$.

Specifically, expressing mn as $a \times b \times c$ or $(a \times b) \times (a+b)$, we have:

$6 = 1 \times 2 \times 3 = (1 \times 2) \times (1+2)$

$30 = 2 \times 3 \times 5 = (2 \times 3) \times (2+3)$

$84 = 3 \times 4 \times 7 = (3 \times 4) \times (3+4)$

$180 = 4 \times 5 \times 9 = (4 \times 5) \times (4+5)$

$330 = 5 \times 6 \times 11 = (5 \times 6) \times (5+6)$

$546 = 6 \times 7 \times 13 = (6 \times 7) \times (6+7)$

Generally, each of the above can be expressed as $a \times b \times c$ or $(a \times b) \times (a+b)$ or as already mentioned above. We can rewrite each of the above as follows:

$1 \times (2 \times 3) = 6 = 1 \times 6$

$2 \times (3 \times 5) = 30 = 5 \times 6$

$3 \times (4 \times 7) = 84 = 14 \times 6$

$4 \times (5 \times 9) = 180 = 30 \times 6$

$5 \times (6 \times 11) = 330 = 55 \times 6$

$6 \times (7 \times 13) = 546 = 91 \times 6$

Each q or mn has 3 and 2 as factors. If this is true, it follows that each q or mn also has 6 as a factor. Therefore,

$1 \times 6 = (1)(2)[(2 \times 1)+1]$

$5 \times 6 = (2)(3)[(2 \times 2)+1]$

$14 \times 6 = (3)(4)[(2 \times 3)+1]$

$30 \times 6 = (4)(5)[(2 \times 4)+1]$

$55 \times 6 = (5)(6)[(2 \times 5)+1]$

$91 \times 6 = (6)(7)[(2 \times 6)+1]$

Remember that the divisibility rule for 6 is a combination of rules for 2 and 3.

From our notation, if the first factor a is n, the following are true:

(a) The second factor is $n + 1$. Since $b - a = 1$, then $b = a + 1$.

By substituting for c, $c = a + b = n + (n+1) = 2n + 1$.

(b) The third factor is c, which is $2n + 1$

(e.g. for $6 \times 7 \times 13$, $n = 6$, $n + 1 = 7$, and $2n + 1 = 13$).

Generally, therefore, each of these products can be represented as:

$(n)(n+1)(2n+1)$.

Also, look at the following:

$(1)(2)(3) = 6 = 6 \times 1$

$(2)(3)(5) = 30 = 6 \times 1 \times 5$

$(3)(4)(7) = 84 = 6 \times 2 \times 7$

$(4)(5)(9) = 180 = 6 \times 5 \times 6$

$(5)(6)(11) = 330 = 6 \times 5 \times 11$

$(6)(7)(13) = 546 = 6 \times 7 \times 13$

$$\frac{1 \times 2 \times 3}{6} = 1,$$

$$\frac{2 \times 3 \times 5}{6} = 5$$

$$\frac{3 \times 4 \times 7}{6} = 14$$

$$\frac{4 \times 5 \times 9}{6} = 30$$

$$\frac{5 \times 6 \times 11}{6} = 55$$

$$\frac{6 \times 7 \times 13}{6} = 91$$

Embedded in each product is the factor 6.

To retain our set of partial sums, we have to divide each of the above products by 6.

Of course, each of these can be represented as:

$$\frac{n(n+1)(2n+1)}{6}.$$

In other words,

$$\sum_{k=1}^{n} k^2 = \frac{n(n+1)(2n+1)}{6}.$$

CHAPTER 5

Another Summation Strategy for First *n* Triangular Numbers

Objectives

At the end of the lesson, the students should be able to:

- Express the sum of first *n* triangular numbers using sigma notation.

- Express in a general form, the sum of squares of *n* consecutive triangular numbers.

Introduction

This chapter explores a strategy for summing first *n* triangular numbers using sigma notation.

It also explores the formula for finding the sum of squares of first *n* triangular numbers.

Partial Sums of First n Triangular Numbers

Look at the following partial sums of first *n* triangular numbers.

$1 = 1$

$1 + 3 = 4$

$1 + 3 + 6 = 10$

$1 + 3 + 6 + 10 = 20$

Paul Emekwulu

$1 + 3 + 6 + 10 + 15 = 35$

$1 + 3 + 6 + 10 + 15 + 21 = 56$

$1 + 3 + 6 + 10 + 15 + 21 + 28 = 84.$

Expressing Partial Sums as Sums of Consecutive Squares

These partial sums can also be represented as:

$1 = 1 + 0 = 1^2 + (0)^2$

$4 = 4 + 0 = 2^2 + (0)^2$

$10 = 9 + 1 = 3^2 + (1)^2$

$20 = 16 + 4 = 4^2 + (2)^2$

$35 = 25 + 10 = 5^2 + (1^2 + 3^2)$

$56 = 36 + 20 = 6^2 + (2^2) + (4)^2$

$84 = 49 + 35 = 7^2 + (1)^2 + (3)^2 + (5)^2$

Study the above to discover the pattern. Derive a formula in terms of n for the sum of first n triangular numbers (impose any restrictions on n if any).

Using your new expression for S_n and taking $n = 8$ for (*a*) and $n = 7$ for (*b*).

show that:

$$S_n = \sum_{k=0}^{\frac{n}{2}} (n - 2k)^2 \text{ or } S_n = \sum_{k=1}^{\frac{n}{2}} (2k)^2$$

for even numbers of first n triangular numbers, and
for odd numbers of first n triangular numbers respectively.

$1 + 3 = 2^2 + (0) = 2^2 + S_0 = S_0 + 2^2$

$1 + 3 + 6 = 3^2 + 1^2 = 3^2 + (1) = 3^2 + S_1 = S_1 + 3^2$

$1 + 3 + 6 + 10 = 4^2 + 2^2 = 4^2 + (1+3) = 4^2 + S_2 = S_2 + 4^2$

$1 + 3 + 6 + 10 + 15 = 5^2 + (1^2+3^2) = 5^2 + (1+3+6) = 5^2 + S_3 = S_3 + 5^2$

Let us draw a table showing the following:

•Number of terms of triangular numbers being summed up in each of the above sums. Call this n.

•Number of terms of triangular numbers equal to the sum enclosed in parenthesis in each case. Call this q.

n	q	$n-q$
3	1	2
4	2	2
5	3	2
6	4	2
7	5	2

Table 17: The Relationship Between n and q

$\sqrt{(1)^2(1)^2} = 1 \times 1 = 1$	$\sqrt{(4)^2(7)^2} = 4 \times 7 = 28$
$\sqrt{(1)^2(3)^2} = 1 \times 3 = 3$	$\sqrt{(4)^2(9)^2} = 4 \times 9 = 36$
$\sqrt{(2)^2(3)^2} = 2 \times 3 = 6$	$\sqrt{(5)^2(9)^2} = 5 \times 9 = 45$
$\sqrt{(2)^2(5)^2} = 2 \times 5 = 10$	$\sqrt{(5)^2(11)^2} = 5 \times 11 = 55$
$\sqrt{(3)^2(5)^2} = 3 \times 5 = 15$	$\sqrt{(6)^2(11)^2} = 6 \times 11 = 66$
$\sqrt{(3)^2(7)^2} = 3 \times 7 = 21$	$\sqrt{(6)^2(13)^2} = 6 \times 13 = 78$

We can show that each of the above square roots is a triangular number by:

(a) Expressing the radicands in terms of n.

(b) Finding the square root.

Next is to express in a general form, the sum (S_n) of squares of n consecutive triangular numbers.

$$1^2 + 3^2 + 6^2 + 10^2 + 15^2 + 21^2 + 28^2 + 36^2 + 45^2 + 55^2 + 66^2 + 78^2 + 91^2 + 105^2$$

We can do this by, first, considering a few individual cases (both even and odd), and then deriving a general form of expressing these products.

Let us express the above products in terms of n.

This investigation involves a pattern exhibited by these examples.

Let us draw a table showing the following:

(i) Number of terms of consecutive triangular numbers being

summed up in each of the above sums. Call this n.

(ii) Number of terms of consecutive triangular numbers equal to the sum enclosed in parenthesis in each case. Call this q.

n	q	$n-q$
3	1	2
4	2	2
5	3	2
6	4	2
7	5	2
8	6	2
9	7	2
10	8	2

Table 18: Relationship between n and q

$q = \{1, 2, 3, 4, 5\}$ and 2, 3, 4, 5, 6, 7... eventually corresponding to:

$$S_1, S_2, S_3, S_4, S_5, S_6$$

We can express the number of terms in q in terms of n.

Each of the numbers in q is 2 less than the number of terms being summed up (i.e. $n - 2 = q$).

Using Pattern Recognition to Find S_n

$2^2, 3^2, 4^2, 5^2, 6^2, 7^2$... correspond to the squares of the number of terms of first n triangular numbers being summed up

(e.g. for $6^2 + S_4$, the numbers being summed up are 1, 3, 6, 10, 15, 21).

Paul Emekwulu

The number of terms being summed up is 6.

For $8^2 + S_6$, the numbers being summed up are as follows:

1, 3, 6, 10, 15, 21, 28, 36.

Therefore, $S_n = S_{n-2} + n^2$.

The above investigation did not include a case when $n = 1$.

Why is this so?

Now, let us look at examples from even cases and then examples from odd cases.

Case 1: When *n* is Even

EXAMPLE 1:
Considering $1 + 3$, $n = 2$.

$S_n = S_{n-2} + n^2$.

By substitution, $S_2 = S_{2-2} + 2^2 = S_0 + 2^2 = 0 + 4 = 4$.

EXAMPLE 2:
Considering $1 + 3 + 6 + 10 + 15 + 21$, $n = 6$.

$S_n = S_{n-2} + n^2$.

By substitution, $S_6 = S_{6-2} + 6^2 = S_4 + 6^2 = 20 + 36 = 56$.

Case 2: When *n* is Odd
EXAMPLE 1:
Considering $1 + 3 + 6 + 10 + 15 + 21 + 28$, $n = 5$.

$S_n = S_{n-2} + n^2$.

$1 + 3 + 6 + 10 + 15 + 21 + 28$, $n = 7$.

$S_n = S_{n-2} + n^2$.

By substitution, $S_7 = S_{7-2} + 7^2 = S_5 + 7^2 = 35 + 49 = 84$.

Sum of Squares of First n Triangular Numbers

In order to differentiate between the sum of $1 + 3 + 6 + 10 + \ldots$ on one hand, and their squares on the other, let our new sum be denoted by S'_n.

In order words:

$S'_n = 1^2 + 3^2 + 4^2 + 6^2 + 10^2 + 15^2 + 21^2 + 28^2 + 36^2 + 45^2 + 55^2 + 66^2 + 78^2 + 91^2 + 105^2 + 120^2 + \ldots$

Using our new notation, we can now make the following statements:

$S'_0 = 0 = 0$.

$S'_1 = 1^2 = 1 = 0 + 1 = (1)^2$.

$S'_2 = 1^2 + 3^2 = 10 = (9 + 1) = (3^2 + 1^2)$.

$S'_3 = 1^2 + 3^2 + 6^2 = 46 = 36 + (9 + 1) = (6^2 + 3^2 + 1^2)$.

$S'_4 = 1^2 + 3^2 + 6^2 + 10^2 = 146 = 100 + (36 + 9 + 1) = (10^2 + 6^2 + 3^2 + 1^2)$.

$S'_5 = 1^2 + 3^2 + 6^2 + 10^2 + 15^2 = 371 = 225 + (100 + 36 + 9 + 1) = (15^2 + 10^2 + 6^2 + 3^2 + 1^2)$.

$S'_1 = S'_0 + (1)^2 (1)^2 = 1$

$S'_2 = S'_1 + (1)^2 (3)^2 = 10$

$S'_3 = S'_2 + (2)^2 (3)^2 = 46$

$S'_4 = S'_3 + (2)^2 (5)^2 = 146$

$S'_5 = S'_4 + (3)^2 (5)^2 = 371$

Paul Emekwulu

Expressing the Square of a Triangular Number A as a Product of Two Squares

The square of any triangular number A, can be expressed as a product of two square numbers a^2 and b^2 where a and b are prime factors of A.

For Odd-Subscripted Triangular numbers

EXAMPLE 1:

When $A = 15$, $15 \times 15 = \left(\dfrac{n+1}{2}\right)^2 (n)^2$

Since $n = 5$, $15 \times 15 = \left(\dfrac{5+1}{2}\right)^2 \times 5^2 = 3^2 \times 5^2$.

EXAMPLE 2:

When $A = 28$, $28 \times 28 = \left(\dfrac{n+1}{2}\right)^2 (n)^2$

Since $n = 7$, $28 \times 28 = \left(\dfrac{7+1}{2}\right)^2 \times 7^2 = 4^2 \times 7^2$.

EXAMPLE3:

When $A = 45$, $45 \times 45 = \left(\dfrac{n+1}{2}\right)^2 (n)^2$

Since $n = 9$, $45 \times 45 = \left(\dfrac{9+1}{2}\right)^2 \times 9^2 = 5^2 \times 9^2$.

EXAMPLE 4:

When $A = 66$, $66 \times 66 = \left(\dfrac{n+1}{2}\right)^2 (n)^2$

Since $n = 11$, $66 \times 66 = \left(\dfrac{11+1}{2}\right)^2 \times (11)^2 = 6^2 \times 11^2$.

.

For Even-Subscripted Triangular Numbers

EXAMPLE 1:

When $A = 21$, $21 \times 21 = \left(\dfrac{n}{2}\right)^2 (n+1)^2$.

Since $n = 6$, $21 \times 21 = \left(\dfrac{6}{2}\right)^2 \times (6+1)^2 = 3^2 \times 7^2$.

EXAMPLE 2:

When $A = 36$, $36 \times 36 = \left(\dfrac{n}{2}\right)^2 (n+1)^2$.

Since $n = 8$, $36 \times 36 = \left(\dfrac{8}{2}\right)^2 \times (8+1)^2 = 4^2 \times 9^2$.

EXAMPLE 1:

When $A = 21$, $21 \times 21 = \left(\dfrac{n}{2}\right)^2 (n+1)^2$.

Since $n = 6$, $21 \times 21 = \left(\dfrac{6}{2}\right)^2 \times (6+1)^2 = 3^2 \times 7^2$.

Paul Emekwulu

EXAMPLE 2 :

When $A = 36$, $36 \times 36 = \left(\dfrac{n}{2}\right)^2 (n+1)^2$.

Since $n = 8$, $36 \times 36 = \left(\dfrac{8}{2}\right)^2 \times (8+1)^2 = 4^2 \times 9^2$.

EXAMPLE 3:

When $A = 55$, $55 \times 55 = \left(\dfrac{n}{2}\right)^2 (n+1)^2$.

Since $n = 10$, $55 \times 55 = \left(\dfrac{10}{2}\right)^2 \times (10+1)^2 = 5^2 \times 11^2$.

EXAMPLE 4:

When $A = 78$, $78 \times 78 = \left(\dfrac{n}{2}\right)^2 (n+1)^2$.

Since $n = 12$, $78 \times 78 = \left(\dfrac{12}{2}\right)^2 \times (12+1)^2 = 6^2 \times 13^2$.

EXAMPLE 5:

When $A = 105$, $105 \times 105 = \left(\dfrac{n}{2}\right)^2 (n+1)^2$.

Since $n = 14$, $105 \times 105 = \left(\dfrac{14}{2}\right)^2 \times (14+1)^2 = 7^2 \times 15^2$.

CHAPTER 6

Exploring Properties of Triangular Numbers

Objectives

At the end of the lesson, the students should be able to:

•List some properties of triangular numbers.

•Use examples to verify these properties.

Introduction

As I discovered a mathematical relationship between the numbers of the Fibonacci sequence and triangular numbers, I started to suspect a mathematical relationship between the two. What do I mean by this relationship? By that I mean a mathematical formula that translates Fibonacci numbers into triangular numbers. This mathematical formula eventually turned out to be a formula that converts any three consecutive Fibonacci numbers a, b, c, into a single, unique triangular number. The following are some properties of triangular numbers.

Property Number 1

The cube of the difference between any two consecutive triangular numbers, a and b, with subscripts, m and n, respectively, is equal to n^3.

Symbolically, for any two consecutive triangular numbers, a and b, with subscripts, m and n, respectively, $(b-a)^3 = n^3$.

Examples with One-Digit Triangular Numbers

EXAMPLE 1:

If $a = 1$, $b = 3$, $m = 1$, $n = 2$.

Paul Emekwulu

$(b–a)^3 = (3–1)^3 = 2^3 = 8$, and $n^3 = 2^3 = 8$.

EXAMPLE 2:

If $a = 3$, $b = 6$, $m = 2$, $n = 3$

$(b-a)^3 = (6-3)^3 = 3^3 = 27$ and $n^3 = 3^3 = 27$.

EXAMPLE 3:

If $a = 15$, $b = 21$, $m = 5$, $n = 6$.

$(b–a)^3 = (21–15)^3 = 6^3 = 216$, and $n^3 = 6^3 = 216$.

EXAMPLE 4:

If $a = 21$, $b = 28$, $m = 6$, $n = 7$.

$(b–a)^3 = (28–21)^3 = 7^3 = 343$, and $n^3 = 7^3 = 343$.

A Search for Counter Examples

Can you identify any two consecutive triangular numbers, a and b, with subscripts, m and n, respectively, where $(b–a)^2$ is not equal to the cube of n?

Property Number 2

The square of any triangular number can be expressed as a product of two squares.

Examples with One-Digit Triangular Numbers

EXAMPLE 1:

$1^2 = 1^2 \times 1^2$

EXAMPLE 2:

$$3^2 = 1^2 \times 3^2$$

EXAMPLE 3:

$$6^2 = 2^2 \times 3^2$$

Examples with Two-Digit Triangular Numbers

EXAMPLE 1:

$$36^2 = 4^2 \times 9^2$$

EXAMPLE 2:

$$55^2 = 5^2 \times 11^2$$

EXAMPLE 3:

$$78^2 = 6^2 \times 13^2$$

A Search for Counter Examples

Can you identify any triangular number whose square cannot be expressed as a product of two squares?

Property Number 3

For any three consecutive Fibonacci numbers, a, b, c, half the product of the sum of subscripts of a and c and the sum of subscripts of a and b is equal to a triangular number whose subscript is equal to the sum of subscripts of a and b.

EXAMPLE 1:

Consider 5, 8, and 13.

$$a = 5 = u_5, b = 8 = u_6, c = 13 = u_7$$

Sum of the subscripts of a and $c = 5 + 7 = 12$.

Call this A.

69

Paul Emekwulu

Sum of the subscripts of a and $b = 5 + 6$.

Call this B.

Half the product of A and $B = \frac{1}{2}(AB) = \frac{1}{2}(11 \times 12) = 66$

and the subscript of 66 is 11, which is the sum of the subscripts

of a and b.

EXAMPLE 2:

Consider 144, 233, and 377.

$$a = 144 = u_{12}, b = 233 = u_{13}, c = 377 = u_{14}$$

Sum of the subscripts of a and $c = 12 + 14 = 26$. Call this A.

Sum of the subscripts of a and $b = 12 + 13 = 25$. Call this B.

Half the product of A and $B = \frac{1}{2}(AB) = \frac{1}{2}(25 \times 26) = 325$

and the subscript of 325 is 25 which is the sum of the subscripts

of a and b.

Property Number 4

For any three consecutive triangular numbers a, b, c,

$$\frac{a \times c}{b} = \frac{n(n+3)}{2},$$

where n is equal to the subscript of a, and

70

$$\frac{n(n+3)}{2}$$

is equal to the number of diagonals in an n-sided polygon.

EXAMPLE 1:

Consider 1, 3, and 6.

$a = 1, b = 3, c = 6.$

$$\frac{a \times c}{b} = \frac{1 \times 6}{3} = 2 \text{ and } \frac{n(n+3)}{2} = \frac{1(1+3)}{2} = 2.$$

EXAMPLE 2:

Consider 3, 6, and 10.

$a = 3, b = 6, c = 10.$

$$\frac{a \times c}{b} = \frac{3 \times 10}{6} = 5 \text{ and } \frac{n(n+3)}{2} = \frac{2(2+3)}{2} = 5$$

EXAMPLE 3:

Consider 6, 10, and 15.

$a = 6, b = 10, c = 15.$

$$\frac{a \times c}{b} = \frac{6 \times 15}{10} = 9 \text{ and } \frac{n(n+3)}{2} = \frac{3(3+3)}{2} = 9.$$

EXAMPLE 4:

Consider 10, 15, and 21.

$a = 10, b = 15, c = 21.$

Paul Emekwulu

$$\frac{a \times c}{b} = \frac{10 \times 21}{15} = 14 \text{ and } \frac{n(n+3)}{2} = \frac{4(4+3)}{2} = 14.$$

A Search for Counter Examples

Can you identify any three consecutive triangular numbers, a, b, c, such that:

$$\frac{a \times c}{b} \neq \frac{n(n+3)}{2} \text{ or } \frac{a \times c}{b} - \frac{n(n+3)}{2} \neq 0,$$

where n is equal to the subscript of a, and

$$\frac{n(n+3)}{2}$$

is equal to the number of diagonals in an n-sided polygon?

Property Number 5

The sum of the number of sides (n) in a three- or more-sided polygon and the number of corresponding diagonal is always equal to a triangular number whose subscript is $n - 1$.

n	Number of diagonals	Corresponding triangular numbers	Subscript
3	0	3	2
4	2	6	3
5	5	10	4
6	9	15	5
7	14	21	6

Table 19: One-to-One Correspondence between Number of Diagonals in an-Sided Polygon and Triangular Numbers

$$n + \frac{n(n-3)}{2} = \frac{2n + n(n-3)}{2}$$

$$= \frac{2n + n^2 - 3n}{2} = \frac{n^2 - n}{2}, \; n \geq 2$$

EXAMPLE 1:

When $n = 3$,

the number of diagonals $= \dfrac{3(3-3)}{2} = 0$.

$T_{n-2} = T_1$, while $1 - 0 = 1$.

EXAMPLE 2:

When $n = 20$,

the number of diagonals $= \dfrac{20(20-3)}{2} = 170$.

$T_{n-2} = T_{18}$ and $T_{18} = 171$, while $171 - 170 = 1$.

A Search for Counter Examples

Can you identify any three- or more-sided polygon whose sum of the number of sides and the number of corresponding diagonal is not always equal to a triangular number whose subscript is $n - 1$?

Property Number 6

For any given n-sided polygon, the number of diagonals is equal to 1 less than a triangular number whose subscript is 2 less than n.

Paul Emekwulu

n	Number of Diagonals (d_n)	n^{th} Triangular Number (t_n)	Subscript (s)	$n - s$	$t_n - d_n$
3	0	1	1	2	1
4	2	3	2	2	1
5	5	6	3	2	1
6	9	10	4	2	1
7	14	15	5	2	1
8	20	21	6	2	1
9	27	28	7	2	1
10	35	36	8	2	1

Table 20: Relationship between Number of Diagonals and Triangular Numbers

EXAMPLE 1:

Let the nth triangular number be T_n.

When $n = 4$, the number of diagonals $= \dfrac{4(4-3)}{2} = 2$.

When $n = 4$, $T_{n-2} = T_2$, and $T_2 = 3$.

$3 - 2 = 1$.

EXAMPLE 2:

Let the nth triangular number be T_n.

When $n = 7$, the number of diagonals $= \dfrac{7(7-3)}{2} = 14$.

When $n = 7$, $T_{n-2} = T_5$ and $T_5 = 15$.

$15 - 14 = 1$.

74

EXAMPLE 3:

Let the nth triangular number be T_n.

When $n = 8$, the number of diagonals $= \dfrac{8(8-3)}{2} = 20$.

When $n = 8$, $T_{n-2} = T_6$ and $T_6 = 21$.

$21 - 20 = 1$.

EXAMPLE 4:

Let the nth triangular number be T_n.

When $n = 10$, the number of diagonals $= \dfrac{10(10-3)}{2} = 35$.

When $n = 10$, $T_{n-2} = T_8$ and $T_8 = 36$.

$36 - 35 = 1$.

A Search for Counter Examples

Can you identify any n-sided polygon whose number of diagonals is not equal to 1 less than a triangular number whose subscript is 2 less than n?

Property Number 7

The sum of squares of any two consecutive triangular numbers, a and b, with subscripts, n and $n + 1$, respectively, is equal to a triangular number, $a^2 + b^2$, whose subscript is equal to $(n+1)^2$.

Examples with One-Digit Triangular Numbers

EXAMPLE 1:

Paul Emekwulu

Consider $a = 1$ and $b = 3$.

Subscript of $a = 1$.

Subscript of $b = 1 + 1 = 2$.

Sum of squares of a and $b = 1^2 + 3^2 = 1 + 9 = 10$.

Subscript of $10 = 4$.

$(n+1)^2 = 2^2 = 4$

EXAMPLE 2:
Consider $a = 3$ and $b = 6$.

Subscript of $a = 2$.

Subscript of $b = 2 + 1 = 3$.

Sum of squares of a and $b = 3^2 + 6^2 = 9 + 36 = 45$.

Subscript of $45 = 9$.

$(n+1)^2 = 3^2 = 9$.

Examples with Two-Digit Triangular Numbers
EXAMPLE 1:
Consider $a = 10$ and $b = 15$.

Subscript of $a = 4$.

Subscript of $b = 4 + 1 = 5$.

Sum of squares of a and $b = 10^2 + 15^2 = 100 + 225 = 325$.

Subscript of $325 = 25$.

$(n+1)^2 = 5^2 = 25$ and this is the subscript of $a^2 + b^2$.

EXAMPLE 2:

Consider $a = 15$ and $b = 21$.

Subscript of $a = 5$.

Subscript of $b = 5 + 1 = 6$.

Sum of squares of a and $b = 15^2 + 21^2 = 225 + 441 = 666$.

Subscript of $666 = 36$.

$(n+1)^2 = 6^2 = 36$ and this is the subscript of $a^2 + b^2$.

A Search for Counter Examples

Can you identify any two consecutive triangular numbers, a and b, with subscripts, n and $n+1$, respectively, is not equal to a triangular number $a^2 + b^2$ whose subscript is equal to $(n+1)^2$?

Property Number 8

For any three consecutive Fibonacci numbers a, b, c,

$$\frac{b^2 + 4ac + (c + a)}{2}$$

is always a triangular number whose subscript is equal to $c + a$.

Examples with One-Digit Fibonacci Numbers

EXAMPLE 1:

Consider 1, 1, and 2.

$a = 1, b = 1, c = 2$.

Paul Emekwulu

$$\frac{b^2 + 4ac + (c+a)}{2} = \frac{1^2 + (4 \times 1 \times 2) + (2+1)}{2} = \frac{1+8+3}{2} = \frac{12}{2} = 6.$$

EXAMPLE 2:

Consider 1, 2, and 3.

$a = 1, b = 2, c = 3.$

$$\frac{b^2 + 4ac + (c+a)}{2} = \frac{2^2 + (4 \times 1 \times 3) + (3+1)}{2} = \frac{4+12+4}{2} = \frac{20}{2} = 10.$$

Examples with Two-Digit Fibonacci Numbers

EXAMPLE 1:

Consider 13, 21, and 34.

$a = 13, b = 21, c = 34.$

$$\frac{b^2 + 4ac + (c+a)}{2} = \frac{21^2 + (4 \times 13 \times 34) + (34+13)}{2}$$
$$= \frac{441 + 1,768 + 47}{2} = \frac{2,256}{2} = 1,128.$$

EXAMPLE 2:

Consider 21, 34, and 55.

$a = 21, b = 34, c = 55.$

$$\frac{b^2 + 4ac + (c+a)}{2} = \frac{34^2 + (4 \times 21 \times 55) + (55+21)}{2}$$
$$= \frac{1,156 + 4,620 + 76}{2} = \frac{5,852}{2} = 2,926.$$

A Search for Counter Examples

Can you identify any three consecutive Fibonacci numbers, a, b, c, such that

$$\frac{b^2 + 4ac + (c + a)}{2}$$

is not always a triangular number whose subscript is not equal to $c + a$?

Property Number 9

For any two consecutive triangular numbers, a and b, with subscripts, x and y, respectively, $x(y+1) = (a+b) - 1$.

Examples with One-Digit Triangular Numbers

EXAMPLE 1:

Consider the numbers 1 and 3.

If $a = 1$ and $b = 3$, then $x = 1$ and $y = 2$.

$x(y+1) = 1(2+1) = 3$

$(a+b) - 1 = (1+3) - 1 = 3$

Therefore, $x(y+1) = (a+b) - 1$.

EXAMPLE 2:

Consider the numbers 3 and 6.

If $a = 3$ and $b = 6$, then $x = 2$ and $y = 3$.

$x(y+1) = 2(3+1) = 8$

$(a+b) - 1 = (3+6) - 1 = 8$

Paul Emekwulu

Therefore, $x(y+1) = (a+b) - 1$.

Examples with Two-Digit Triangular Numbers

EXAMPLE 1:

Consider the numbers 36 and 45.

If $a = 36$ and $b = 45$, then $x = 8$ and $y = 9$.

$x(y+1) = 8(9+1) = 80$

$(a+b) - 1 = (36+45) - 1 = 80$

Therefore, $x(y+1) = (a + b) - 1$.

EXAMPLE 2:

Consider the numbers 91 and 105.

If $a = 91$ and $b = 105$, then $x = 13$ and $y = 14$.

$x(y+1) = 13(14+1) = 195$

$(a+b) - 1 = (91+105) - 1 = 195$

Therefore, $x(y+1) = (a+b) - 1$.

A Search for Counter Examples

Can you identify any two consecutive triangular numbers, a and b, with subscripts, x and y, respectively, such that $x(y+1) \neq (a+b) - 1$?

Property Number 10

For any two consecutive triangular numbers, a and b,

$$.a - b = \frac{a+b}{a-b} \text{ or } b - a = \frac{a+b}{b-a}.$$

Examples with One-Digit Triangular Numbers

EXAMPLE 1:

Consider 1 and 3.

When $a = 1$ and $b = 3$,

$a - b = 1 - 3 = -2$ and $a + b = 4$, so

$$\frac{a+b}{a-b} = \frac{4}{-2} = -2.$$

$b - a = 3 - 1 = 2$

$a + b = 4$, so $\dfrac{a+b}{b-a} = \dfrac{4}{2} = 2$

Therefore, for any two consecutive triangular numbers, a and b,

$$.a - b = \frac{a+b}{a-b} \text{ or } b - a = \frac{a+b}{b-a}.$$

EXAMPLE 2:

Consider 3 and 6.

When $a = 3$ and $b = 6$,

$a - b = 3 - 6 = -3$

and $a + b = 9$, so $\dfrac{a+b}{a-b} = \dfrac{9}{-3} . = -3$

Paul Emekwulu

$b - a = 6 - 3 = 3$ and $a + b = 3 + 6 = 9$, so

$$\frac{a+b}{b-a} = \frac{9}{3} = 3$$

Therefore, for any two consecutive triangular numbers, a and b,

$$a - b = \frac{a+b}{a-b} \text{ or } b - a = \frac{a+b}{b-a}.$$

Examples with Two-Digit Triangular Numbers

EXAMPLE 1:

Consider 15 and 21.

When $a = 15$ and $b = 21$,

$a - b = 15 - 21 = -6$ and $a + b = 36$, so

$$\frac{a+b}{a-b} = \frac{36}{-6} = -6$$

$b - a = 21 - 15 = 6$

$a + b = 15 + 21 = 36$, so

$$\frac{a+b}{b-a} = \frac{36}{6} = 6$$

Therefore, for any two consecutive triangular numbers, a and b,

$$a - b = \frac{a+b}{a-b} \text{ or } b - a = \frac{a+b}{b-a}.$$

EXAMPLE 2:

Consider 28 and 36.

When $a = 28$ and $b = 36$,

$a - b = -8$ and $a + b = 64$, so

$b - a = 36 - 28 = 8$

$$\frac{a+b}{b-a} = \frac{28+36}{36-28} = \frac{64}{8} = 8$$

$b - a = 36 - 28 = 8$, so

$$\frac{a+b}{b-a} = \frac{28+36}{36-28} = \frac{64}{8} = 8$$

Therefore, for any two consecutive triangular numbers, a and b,

$$a - b = \frac{a+b}{a-b} \text{ or } b - a = \frac{a+b}{b-a}.$$

A Search for Counter Examples

Can you identify any two consecutive triangular numbers, a and b, such that

$$a - b \neq \frac{a+b}{a-b} \text{ or } b - a \neq \frac{a+b}{b-a}?$$

Property Number 11

For any two consecutive triangular numbers, a and b,

$$b^2 - a^2 = (b-a)^3 = \sqrt{(a+b)^3}$$

Paul Emekwulu

Examples with One-Digit Triangular Numbers

EXAMPLE 1:

Consider 1 and 3.

$b^2 - a^2 = 3^2 - 1^2 = 9 - 1 = 8$

$(b-a)^3 = (3-1)^3 = 2^3 = 8$

$\sqrt{(a+b)^3} = \sqrt{(1+3)^3} = \sqrt{64} = 8$

EXAMPLE 2:

Consider 3 and 6.

$b^2 - a^2 = 6^2 - 3^2 = 36 - 9 = 27$

$(b-a)^3 = (6-3)^3 = 3^3 = 27$

$\sqrt{(a+b)^3} = \sqrt{(3+6)^3} = \sqrt{729} = 27$

Examples with Two-Digit Triangular Numbers

EXAMPLE 1:

Consider 10 and 15.

$b^2 - a^2 = 15^2 - 10^2 = 225 - 100 = 125$

$(b-a)^3 = (15-10)^3 = 5^3 = 125$

$\sqrt{(a+b)^3} = \sqrt{(10+15)^3} = \sqrt{15625} = 125$

84

EXAMPLE 2:

Consider 21 and 28.

$b^2 - a^2 = 28^2 - 21^2 = 784 - 441 = 343$

$(b-a)^3 = (28-21)^3 = 7^3 = 343$

$\sqrt{(a+b)^3} = \sqrt{(21+28)^3} = \sqrt{117649} = 343$

A Search for Counter Examples

Can you identify any two consecutive triangular numbers, a and b, such that

$b^2 - a^2 \neq (b-a)^3 \neq \sqrt{(a+b)^3}$?

Property Number 12

For any two consecutive triangular numbers, a and b, with subscripts, x and y, respectively,

$$\frac{(x+y)+1}{2} = y$$

Examples with One-Digit Triangular Numbers

EXAMPLE 1:

Let $a = 1$ and $b = 3$.

If $a = 1$, $x = 1$, and if $b = 3$, then $y = 2$.

85

Paul Emekwulu

By substitution,

$$\frac{(x+y)+1}{2} = \frac{(1+2)+1}{2} = 2$$

Therefore, $\frac{(x+y)+1}{2} = y.$

EXAMPLE 2:
Let $a = 3$ and $b = 6$.

If $a = 3$, $x = 2$, and $b = 6$, then $y = 3$.

By substitution,

$$\frac{(x+y)+1}{2} = \frac{(2+3)+1}{2} = 3$$

Therefore, $\frac{(x+y)+1}{2} = y.$

Examples with Two-Digit Triangular Numbers

EXAMPLE 1:
Let $a = 10$ and $b = 15$.

If $a = 10$, $x = 4$, and $b = 15$, then $y = 5$.

By substitution,

$$\frac{(x+y)+1}{2} = \frac{(4+5)+1}{2} = 5$$

Therefore, $\dfrac{(x+y)+1}{2} = y$.

EXAMPLE 2:

Let $a = 45$ and $b = 55$.

If $a = 45$, $x = 9$, and $b = 55$, then $y = 10$.

By substitution,

$$\frac{(x+y)+1}{2} = \frac{(9+10)+1}{2} = 10$$

Therefore, $\dfrac{(x+y)+1}{2} = y$.

A Search for Counter Examples

Can you identify any two consecutive triangular numbers a and b such that

$$\frac{(x+y)+1}{2} \neq y?$$

Property Number 13

For any three consecutive Fibonacci numbers, $a, b, c,$

$$\frac{b^2 + 4ac - (c+a)}{2}$$

is always a triangular number whose subscript is equal to $c + a$.

Paul Emekwulu

Examples with One-Digit Fibonacci Numbers

EXAMPLE 1:

Consider 1, 1, and 2.

$a = 1, b = 1, c = 2$.

$$\frac{b^2 + 4ac - (c+a)}{2} = \frac{1^2 + (4 \times 1 \times 2) - (2+1)}{2} = \frac{1+8-3}{2} = \frac{6}{2} = 3.$$

EXAMPLE 2:

Consider 1, 2, and 3.

$a = 1, b = 2, c = 3$.

$$\frac{b^2 + 4ac - (c+a)}{2} = \frac{2^2 + (4 \times 1 \times 3) - (3+1)}{2} = \frac{4+12-4}{2} = \frac{12}{2} = 6.$$

Examples with Two-Digit Fibonacci Numbers

EXAMPLE 1:

Consider 13, 21, and 34.

$a = 13, b = 21, c = 34$.

$$\frac{b^2 + 4ac - (c+a)}{2} = \frac{21^2 + (4 \times 13 \times 34) - (34+13)}{2}$$
$$= \frac{441 + 1,768 - 47}{2} = \frac{2,162}{2} = 1,081.$$

EXAMPLE 2:

Consider 21, 34, and 55.

$a = 21, b = 34, c = 55$.

$$\frac{b^2 + 4ac - (c+a)}{2} = \frac{34^2 + (4 \times 21 \times 55) - (55 + 21)}{2}$$

$$= \frac{1,156 + 4,620 - 76}{2} = \frac{5,700}{2} = 2,850.$$

A Search for Counter Examples

Can you identify any three consecutive Fibonacci numbers, a, b, c, such that

$$\frac{b^2 + 4ac - (c+a)}{2}$$

is not always a triangular number whose subscript is not equal to $c + a$?

CHAPTER 7

Exploring Independent Trial Questions

1. Find two consecutive triangular numbers whose sum is 144.

2. The sum of three consecutive triangular numbers, x, y, z, is 31. Find the numbers.

3. Show that

$$\frac{n(n+1)}{2}$$

represents a triangular number if, and only if, $n \geq 1$.

4. Prove that the difference between two consecutive triangular numbers, a and b, is equal to the subscript of b.

n	k	$2k-1$
1	1	1
3	2	3
5	3	5
7	4	7
9	5	9
11	6	11
13	7	13
15	8	15

Table 20: General Form o f an Odd Number

3. Given that

$$\sqrt{b^2 + 4ac} = c + a,$$

where a, b, c are consecutive Fibonacci numbers, prove that

$b = c - a$

n	k	2k– 1
1	1	1
3	2	3
5	3	5
7	4	7
9	5	9
11	6	11
13	7	13
15	8	15

n	k	2k– 1
1	1	1
3	2	3
5	3	5
7	4	7
9	5	9
11	6	11
13	7	13
15	8	15

6. In Table **20**, find k in terms of n. Prove that $k(2k–1)$ is a triangular number.

7. Study the table below. Find the pattern and express y in terms of x.

y	x	y–x
2	2	0
4	3	1
6	4	2
8	5	3
10	6	4
12	7	5
14	8	6
16	9	7

8. Using the property of one $[\frac{a}{a}=1, (a \neq 0)]$, show that:

$$\frac{(n)(n+1)(n+2)}{6} = \frac{(n+2)}{6(n-1)!}.$$

9. The difference between two consecutive triangular numbers is 15. Find the numbers using simultaneous equations.

10.

Study the information below:

$1 = 1 = 2^0$

$1 + 1 = 2 = 2^1$

$1 + 2 + 1 = 4 = 2^2$

$1 + 3 + 3 + 1 = 8 = 2^3$

$1 + 4 + 6 + 4 + 1 = 16 = 2^4$

$1 + 5 + 10 + 10 + 5 + 1 = 32 = 2^5$

$1 + 6 + 15 + 20 + 15 + 6 + 1 = 64 = 2^6$

$1 + 7 + 21 + 35 + 35 + 21 + 7 + 1 = 128 = 2^7$

Each sum has been expressed in the form,

2^q, where $q \in W$.

Now complete the following two exercises:

(a) If n represents the number of terms being summed up, using a table, find q in terms of n.

(b) Use the relationship in exercise (a) above to find S_n in terms of n.

11. If a triangular number can be expressed as:

$$\frac{(n+1)!}{2(n-1)!}$$

Complete the following two exercises:

(a) Find this expression in terms of k if $n = 2k - 1$.

(b) Using your new expression from exercise (a) above:

(i) Find the first six triangular numbers.

(ii) What do you notice?

12. Study the table below.

a	x_1	b	x_2	c	x_3
1	1	3	2	6	3
3	2	6	3	10	4
6	3	10	4	15	5
10	4	15	5	21	6
15	5	21	6	28	7
21	6	28	7	36	8
28	7	36	8	45	9

Table 21

Now answer these questions:
(i) For each a, b, c, in the above table, find

$x_1 + x_3$ and $\dfrac{x_1 + x_3}{2}$, where x_1, x_2, x_3

93

are subscripts of a, b, c, respectively (see above table).

(ii) What do you notice between x_2 and $\dfrac{x_1 + x_3}{2}$?

13. Given that $b - a = x_2$, prove that

$$\frac{x_1 + x_3}{2} = x_2$$

where a, b, c are three consecutive triangular numbers with x_1, x_2, x_3 as their corresponding subscripts (see   Table 21).

14. If $2n^2 - 3n + 1,\ n \geq 2$

represents triangular numbers of even subscripts, while

$2n^2 + 3n + 1,\ n \geq 0$

represents triangular numbers of odd subscripts, show that the difference between the first odd-subscripted and the first even-subscripted triangular numbers are even numbers greater than or equal to 2.

Challenge Questions

1. Prove that each of the following represents a triangular number:

(i) $(n-1)(2n-1),\ n \geq 2$

(ii) $\dfrac{(n+1)!}{2(n-1)!},\ n \geq 1$

2. Prove that for any two consecutive triangular numbers a and b,

$b^2 - a^2 = (b - a)^3$.

94

3. Prove that for any two consecutive triangular numbers, a and b, with subscripts, n and

$n+1$, respectively, $b^2 - a^2 = (n+1)^3$.

4. Show that $n(2n+1)$ and $2n^2 + 3n + 1$, $n \geq 2$ represent triangular numbers of odd subscripts.

5. Prove that for any three consecutive Fibonacci numbers, a, b, c,

$$\frac{b^2 + 4ac - (c + a)}{2}$$

is always a triangular number.

6.

$S_1 = 1$

$S_2 = 1 + 3 = 4$

$S_3 = 1 + 3 + 6 = 10$

$S_4 = 1 + 3 + 6 + 10 = 20$

$S_5 = 1 + 3 + 6 + 10 + 15 = 35$

$S_6 = 1 + 3 + 6 + 10 + 15 + 21 = 56$

$S_7 = 1 + 3 + 6 + 10 + 15 + 21 + 28 = 84$

From the above, the following is true:

$S_0 + S_1 = 1$

$S_1 + S_2 = 1 + 4 = 5$

Paul Emekwulu

$S_2 + S_3 = 4 + 10 = 14$

$S_3 + S_4 = 10 + 20 = 30$

$S_4 + S_5 = 20 + 35 = 55$

$S_5 + S_6 = 35 + 56 = 91$

$S_6 + S_7 = 56 + 84 = 140$

If $S_n = \dfrac{n(n+1)(n+2)}{6}$,

prove that $S_n + S_{n-1} = \dfrac{n(n+1)(2n+1)}{6}$,

7. The difference between two consecutive triangular numbers is 5. Find the numbers using simultaneous equations.

8. For any two consecutive triangular numbers, x and y, the sum is equal to the square of their difference. Derive the fact that

$$\frac{(y-x)(y-x)}{y+x} = 1.$$

9. The square of the difference between two consecutive triangular numbers, x and y, is 343. Find the numbers.

10.

n	d_n	t_n	t_n-d_n	Subscript for t_n
3	0	3	3	2
4	2	...	...	...
5	5	...	...	...
6	9	...	...	...
7	14	...	...	...
8	20	...	...	...
9	27	...	...	...
10	35	...	...	...
11	44	55	...	...

n = the number of sides of a polygon

d_n = the number of diagonals

t_n = the corresponding triangular numbers

Study the above table, fill in the missing information, and then complete the following statement:

For any given n – sided polygon, the number of diagonals is equal to 1 less than a triangular number whose subscript

11. Study the first partial sums below whose first term is equal to 2. Prove that the sum of first n terms is given by:

$$\frac{n(n+3)}{2}$$

$S_1 = 2 = 2$

$S_2 = 2 + 3 = 5$

$S_3 = 2 + 3 + 4 = 9$

Paul Emekwulu

$S_4 = 2 + 3 + 4 + 5 = 14$

$S_5 = 2 + 3 + 4 + 5 + 6 = 20$

$S_6 = 2 + 3 + 4 + 5 + 6 + 7 = 27$

$S_7 = 2 + 3 + 4 + 5 + 6 + 7 + 8 = 35$

$S_8 = 2 + 3 + 4 + 5 + 6 + 7 + 8 + 9 = 44$

$S_9 = 2 + 3 + 4 + 5 + 6 + 7 + 8 + 9 + 10 = 54$

$S_{10} = 2 + 3 + 4 + 5 + 6 + 7 + 8 + 9 + 10 + 11 = 65$

12.

a	b	c	$\dfrac{(c+a)^2 - (c+a)}{2}$	Subscript
1	1	2	3	...
1	2	3	6	...
2	3	5	21	...
3	5	8	55	...
5	8	13	153	...
8	13	21	406	...
13	21	34	...	...
21	34	55	...	...
34	55	89	...	...
55	89	144	...	...

For any three consecutive Fibonacci numbers, a, b, c,

$$\frac{(c+a)^2 - (c+a)}{2}$$

is a triangular number (see above table).

Do the following exercises:

(a) In a general form, express the subscripts of your results in column 4 in terms of Fibonacci numbers.

(b) Fill in the missing information in column 5.

(c) Find the difference between consecutive terms in column 5.

13. Study Table 22 below and answer the following question:

Given the fact that a, b, c are consecutive triangular numbers with subscripts, n,

$n + 1$, and $n + 2$, respectively, prove that:

$$\frac{a \times c}{b} = \frac{n(n-3)}{2}, \; n \geq 1.$$

$\dfrac{1 \times 6}{3} - \dfrac{(1 \times 1)(2 \times 3)}{1 \times 3} - 2$	$\dfrac{6 \times 15}{10} - \dfrac{(2 \times 3)(3 \times 5)}{2 \times 5} - 9$	$\dfrac{15 \times 28}{21} - \dfrac{(3 \times 5)(4 \times 7)}{3 \times 7} - 20$
$\dfrac{3 \times 10}{6} - \dfrac{(1 \times 3)(2 \times 5)}{2 \times 3} - 5$	$\dfrac{10 \times 21}{15} - \dfrac{(2 \times 5)(3 \times 7)}{3 \times 5} - 14$	$\dfrac{21 \times 36}{28} - \dfrac{(3 \times 7)(4 \times 9)}{4 \times 7} - 27$

Table 22: Number of Diagonals and Triangular Numbers

CHAPTER 8

Solutions to Independent Trial Questions

1. For any two consecutive triangular numbers, a and b, the sum of a and b is equal to the square of their difference. In other words, $a + b = (b{-}a)^2$. Therefore,

$a + b = 144$(i)

$b - a = 12$(ii)

From equations (i) and (ii), by addition, $2b = 156$.

From here, $b = 78$.

If in equation (ii), $b - a = 12$, then by substitution, $78 - a = 12$(iii).

From equation (iii), $a = 78 - 12 = 66$.

Therefore, the two consecutive triangular numbers are 66 and 78.

2. Let the consecutive triangular numbers be:

$$\frac{n(n+1)}{2}; \frac{(n+1)(n+2)}{2}; \frac{(n+2)(n+3)}{2}$$

Adding these numbers, we have:

$$\frac{n(n+1)}{2} + \frac{(n+1)(n+2)}{2} + \frac{(n+2)(n+3)}{2} = 31$$

$$\frac{\left(n^2 + n\right) + \left(n^2 + 3n + 2\right) + \left(n^2 + 5n + 6\right)}{2} = 31$$

$$\frac{\left(n^2 + n^2 + n^2\right) + \left(n + 3n + 5n\right) + (2 + 6)}{2} = 31$$

$$\frac{3n^2 + 9n + 8}{2} = 31$$

By cross-multiplication, we have:

$3n^2 + 9n + 8 = 62$

Subtracting 8 from both sides of the equation, we have:

$3n^2 + 9n - 54 = 0$.

Dividing both sides of the equation by 3, we have:

$n^2 + 3n + 18 = 0$

Factoring, we have:

$(n-3)(n+6) = 0$

From here, $n = 3$.

If $n = 3$, by substitution,

$$\frac{n(n+1)}{2} = 6,$$

$$\frac{(n+1)(n+2)}{2} = 10,$$

$$\frac{(n+2)(n+3)}{2} = 15.$$

The numbers are 6, 10, and 15.

3. Since the smallest triangular number is 1,

$\dfrac{n(n+1)}{2} \geq 1$ implies $n(n+1) \geq 2$ implies $n^2 + n - 2 \geq 0$

$(n+2)(n-1) \geq 0$

Either $n + 2 \geq 0$ or $n - 1 \geq 0$.

If $n + 2 \geq 0$, then $n \geq -2$.

If $n - 1 \geq 0$, then $n \geq 1$.

Therefore, for $\dfrac{n(n+1))}{2}$ to represent a triangular number, the condition , $n \geq 1$ must exist.

4. If $a = \dfrac{n(n+1)}{2}$, then $b = \dfrac{(n+1)(n+2)}{2}$.

If the subscript of a is n, then the subscript of b is $n + 1$.

$\dfrac{(n+1)(n+2)}{2} - \dfrac{n(n+1)}{2}$

$= \dfrac{n^2 + 3n + 2 - \left(n^2 + n\right)}{2}$

$= \dfrac{n^2 + 3n + 2 - n^2 - n}{2}$

$= \dfrac{2n+2}{2} = \dfrac{2(n+1)}{2} = n+1$

Therefore, the difference between two consecutive triangular numbers, a and b, is equal to the subscript of b.

5.

$$\sqrt{b^2 + 4ac} = c + a$$

Taking the square on both sides, we have:

$$\left(\sqrt{b^2 + 4ac}\right)^2 = (c + a)^2$$

$$b^2 + 4ac = c^2 + 2ac + a^2$$

$$b^2 = c^2 + 2ac + a^2 - 4ac = c^2 - 2ac + a^2 = (c - a)^2$$

Taking the square on both sides we have:

$$b = c - a.$$

6.

n	k	$2k-1$
1	1	1
3	2	3
5	3	5
7	4	7
9	5	9
11	6	11
13	7	13
15	8	15

From the table above,

$$k = \frac{n+1}{2}.$$

Paul Emekwulu

By substitution,

$$k(2k-1) = \left(\frac{n+1}{2}\right)\left[2\left(\frac{n+1}{2}\right)-1\right]$$

$$= \left(\frac{n+1}{2}\right)(n+1-1)$$

$$= \frac{n(n+1)}{2}.$$

and this is the general form of the nth triangular number.

Therefore, $k(2k-1)$ represents a triangular number.

7.

y	x	$y - x$
2	2	0
4	3	1
6	4	2
8	5	3
10	6	4
12	7	5
14	8	6
16	9	7

By inspection of the table above, $2[(y-x)+1] = y$.

$2[(y-x)+1] = y$

$2y - 2x + 2 = y$

$2y - y = 2x - 2$

From here, $y = 2x - 2$.

104

Answer: $y = 2x - 2$.

8.

$$\frac{(n+2)!}{6(n-1)!} = \frac{(n+2)(n+1)(n)(n-1)!}{6(n-1)!}$$

$$= \frac{n(n+1)(n+2)}{6}$$

Therefore, $\frac{n(n+1)(n+2)}{6} = \frac{(n+2)!}{6(n-1)!}$.

9. Let the two consecutive triangular numbers be x and y.

Therefore, $y - x = 15$ (i).

Also, the sum of two consecutive triangular numbers, x and y, is equal to the square of their difference.

i.e. $(y-x)^2 = y + x$.

By substitution,

$y + x = (15)^2 = 225$ (ii).

Combining Equations (i) and (ii), we have:

$y - x = 15$

$y + x = 225$

By addition, $2y = 240$.

Paul Emekwulu

From here, $y = 120$.

From Equation (i), $x = y - 15 = 120 - 15 = 105$.

Therefore, $x = 105$ and $y = 120$.

Answers: 105 and 120.

10(a)

n	q	$n - q$	$n + q$
1	0	1	1
2	1	1	3
3	2	1	5
4	3	1	7
5	4	1	9
6	5	1	11
7	6	1	13

From the above table, $n - q = 1$.

From here, $q = n - 1$.

OR

$n + q = 2n - 1$

Subtracting n from both sides we have:

$(n + q) - n = 2n - 1 - n$

From here, $q = n - 1$.

(b) $S_n = 2^{n-1}$.

11(a) Let $\dfrac{(n+1)!}{2(n-1)!} = Q$.

If $n = 2k - 1$, then

$$Q = \frac{(2k-1+1)!}{2(2k-1-1)!}$$

$$= \frac{(2k)!}{2(2k-2)!} = \frac{(2k)(2k-1)(2k-2)!}{2(2k-2)!}$$

$$= k(2k-1)$$

Answers: (a) $\dfrac{(n+1)!}{2(n-1)!} = k(2k-1)$.

(b) (i)

When $k = 1$, $k(2k-1) = 1[(2\times1)-1] = 1$.

When $k = 2$, $k(2k-1) = 2[(2\times2)-1] = 6$.

When $k = 3$, $k(2k-1) = 3[(2\times3)-1] = 15$.

When $k = 4$, $k(2k-1) = 4[(2\times4)-1] = 28$.

When $k = 5$, $k(2k-1) = 5[(2\times5)-1] = 45$.

When $k = 6$, $k(2k-1) = 6[(2\times6)-1] = 66$.

(ii) 1, 6, 15, 28, 45, 66 are all odd-subscripted.

12(i)

Paul Emekwulu

a	b	c	x_1	x_2	x_3	$x_1 + x_3$	$\dfrac{x_1 + x_3}{2}$
1	3	6	1	2	3	4	2
3	6	10	2	3	4	6	3
6	10	15	3	4	5	8	4
10	15	21	4	5	6	10	5
15	21	28	5	6	7	12	6
21	28	36	6	7	8	14	7

Table 23

(ii) $x_2 = \dfrac{x_1 + x_3}{2}$.

13. Prove that $\dfrac{x_1 + x_3}{2} = x_2$.

Proof:

$$\frac{x_1 + x_3}{2} = \frac{\left(\sqrt{a+b} - 1\right) + \left(\sqrt{a+b} + 1\right)}{2}$$

$$= \frac{(b - a - 1) + (b - a + 1)}{2}$$

$$= \frac{(b + b) - (a + a) - (1 - 1)}{2}$$

$$\frac{2b - 2a}{2} = \frac{2(b - a)}{2} = b - a$$

But $\sqrt{a+b} = b - a = x_2$

Therefore, $\dfrac{x_1 + x_3}{2} = x_2$

Since $\dfrac{x_1 + x_3}{2} = b-a$ and $b-a = x_2,$

$\dfrac{x_1 + x_3}{2} = x_2 .$

14. Let the n in $2n^2 - 3n + 1$ be n', while the n in $2n^2 + 3n + 1$ remains unchanged.

Tabulate values of n and n' for which $2n'^2 - 3n' + 1$ and

$2n^2 + 3n + 1$ have different numerical values.

n'	$2n'^2 - 3n' + 1$	n	$2n^2 + 3n + 1$
2	3	0	1
3	10	1	6
4	21	2	15
5	36	3	28
6	55	4	45
7	78	5	66
8	105	6	91

Find the difference and the relationship between n and n'.

n'	n	$n' - n$
2	0	2
3	1	2
4	2	2
5	3	2
6	4	2
7	5	2
8	6	2

Paul Emekwulu

Table 24: Relationship between n and n'.

From Table 24, we have:

$$\frac{n'+n-2}{2} = n \text{ implies } n'+n-2 = 2n.$$

From here, $n' = n + 2$.

Now substituting for n' in $2n'^2 - 3n' + 1$, we have:

$$2(n+2)^2 - 3(n+2) + 1 = 2(n^2 + 4n + 4) - (3n+6) + 1$$

$$= 2n^2 + 8n + 8 - 3n - 6 + 1$$

$$= 2n^2 + 5n + 3.$$

Subtracting $2n^2 + 3n + 1$ from $2n^2 + 5n + 3$, we have:

$(2n^2+5n+3) - (2n^2+3n+1) = (2n^2 - 2n^2) + (5n-3n) + (3-1) = 2n + 2, n \geq 0.$

Appendix A

Set of Numbers	Summation Formula
1, 3, 5, 7, 9, 11, 13, 15, 17, 19, 21, 23, 25, 27, 29...	n^2
0, 1, 2, 3, 4, 5, 6, 7, 8, 9, 10, 11, 12, 13, 15, 17, 19...	$\dfrac{n}{2}(n-1)$
0, 2, 4, 6, 8, 10, 12, 14, 16, 18, 20, 22, 24, 26, 28, 30...	$n^2 - n$
1, 2, 3, 4, 5, 6, 7, 8, 9, 10, 11, 12, 13, 14, 15, 17...	$\dfrac{n}{2}(a+L),\ \dfrac{n}{2}(n+1),$ $\dfrac{n}{2}\big[2a+(n-1)d\big]$
2, 4, 6, 8, 10, 12, 14, 16, 18, 20...	$n(a+n-1)$
3, 5, 7, 9, 11, 13, 15, 17, 19, 21, 23, 25, 27, 29, 31, 33, 35, 37, 39, 41...	$n^2 + 2n$

Table 27: Different Summation Strategies.

When n is Even	When n is Odd	Either Odd or Even
$\displaystyle\sum_{k=1}^{\frac{n+2}{2}}(2k-2)^2$	$\displaystyle\sum_{k=0}^{\frac{n-1}{2}}(2k+1)^2$	$\dfrac{1}{6}(n)(n+1)(n+2)$
$\displaystyle\sum_{k=0}^{\frac{n-2}{2}}(2k+2)^2$	$\displaystyle\sum_{k=1}^{\frac{n+1}{2}}(2k-1)^2$	$S_{n-1} + \dfrac{(n+1)!}{2(n-1)!}$
$\displaystyle\sum_{k=0}^{\frac{n-2}{2}}(n-2k)^2$	$\displaystyle\sum_{k=0}^{\frac{n-1}{2}}(n-2k)^2$	$S_n = S_{n-2} + n^2$

Table 28: Summation Formulas for First n Triangular Numbers

Paul Emekwulu

Seminars & Seminar Scheduling

Math—Magic **with Paul Chika Emekwulu**

Program Objectives

- Develops analytical and logical thinking skills.

- Presents the beauty, elegance, and excitement in number concepts.

- Entertains and stimulates interest through number tricks and investigatory lessons.

- Encourages pattern recognition.

- Actively involves and motivates students of all abilities.

- Encourages student-teacher, teacher-student, student-student communication.

- Supports the National Council of Teachers of Mathematics (NCTM) standards.

Program Description

Math—Magic is neither about magic nor numerology. Math—Magic is a program that uses creative and innovative teaching strategies to make mathematics exciting, interesting, and

112

intriguing to high school students using paper and pencil. Most of these activities are embedded in guided discovery lessons that come in worksheet format.

Math—Magic is about motivation. It is about excitement. It is about mathematical reasoning. It is about pattern recognition. It is not about magicians' magic. It is not about numerology.

Past Engagements

Math—Magic has been presented to the following schools and associations:

- Oklahoma Council of Teachers of Mathematics

- Panhandle Mathematics & Science Conference

- Kansas Association of Teachers of Mathematics

- Booker T. Washington High School, Tulsa, OK

- Oklahoma City Community College, Oklahoma City,

- Tecumseh High School, Tecumseh, OK

- Jenks Public Schools, Jenks, OK

- Newkirk High School, Newkirk, OK

- Norman High School, Norman, OK

- Oklahoma State University (OSU), Stillwater

- Washington High School, Washington, OK

- Oklahoma Education Association

- Liberated Arts Center, Oklahoma City

- Oklahoma State University (Upward Bound)

- National Council of Teachers of Mathematics

- Board Members of Organization of Rural Oklahoma Schools

114

What People are Saying

"The looks on the students' faces were priceless and the 'Ah-Has' were abundant when the same equations and theories were introduced in new and interesting ways. We would recommend it to anyone who finds it difficult to grasp math concepts. 'Math—Magic' may be the key to unlock the mysteries of the math world."

Bennie Boykin
Upward Bound Director
Oklahoma State University
(Technical Branch) Oklahoma City

"The Central Regional Conference Meeting of the National Council of Teachers of Mathematics in Topeka, KS will be called a resounding success because of people such as you who gave so generously of their time and effort. We are convinced that our conference was among the very best of regional conferences that NCTM has had."

Dr Connie S Schrock
Co-Program Chair
NCTM Central Regional Conference, Topeka, KS

National Council of Teachers of Mathematics
Central Regional Conference
St. Louis, Missouri

29–31 January, 1998

Dear Paul,

What do 400 excellent speakers, nearly 2400 participants, top-notch facilities, well-orchestrated arrangements, spring-like

Paul Emekwulu

weather, and 'show-me' hospitality equal? A memorable Saint Louis Central Regional Conference of the National Council of Teachers of Mathematics!

The accolades for the quality of the program and local arrangements are still arriving. Your presentation contributed to the praise the Program Committee continues to receive. We appreciate the time, thought, and preparation you gave to your part in the great success of our conference.

Best Wishes and many, many thanks,

Sincerely,

Carol A. Edwards

Program Chair

Organization of Rural Oklahoma Schools
Box 189
Foss, OK 73647

March 1, 1995

Paul Chika Emekwulu
Novelty Books
P.OBox 2482
Norman, OK 73070

Dear Paul,
Thank you for your participation during the January meeting of the OROS board of directors.

The group was very pleased with your presentation. I will be pleased to present members of the OROS board with copies of the "Program Request Form."

It is quite evident that you are able to capture the attention of your audience through "Magic with Numbers."

Sincerely,

Tom Butler

Executive Director, OROS

The University of Oklahoma
CENTER FOR THE STUDY OF SMALL RURAL SCHOOLS
COLLEGE OF CONTINUING EDUCATION

February 10, 1997

Mr. Paul Chika Emekwulu
Publisher
Novelty Books
P.O.Box 2482
Norman, OK 73070

Dear Mr. Emekwulu,

Congratulations! Your presentation entitled Math—Magic: Encouraging Mathematical Reasoning, Achieving Motivation and Excitement in the Classroom for the sixth annual National Conference on Creating the Quality School has been reviewed and accepted. The response to the call for presenters has been gratifying. The conference should prove to be exciting and valuable to attendees and presenters alike. The proposals were reviewed as quickly as possible by a panel to enable presenters ample notification to make travel plans. The conference begins Thursday, March 20 and concludes Saturday, March 22 at 10.30 a.m. (see enclosure).
A full registration brochure is included. Remember that registration is required for all presenters. Your pre-registration will help speed up the check-in process at the conference. To register by phone, call: 1-800-527-0772 ext. 2248. If you will not be able

Paul Emekwulu

to attend, please notify us immediately so that we may allow someone else the opportunity to present. If you bring handouts, prepare for approximately 20-25 participants in your session. Please make your session as interactive as possible.

Please be sure your name, the name of your organization, and the address at the top of this letter are correct and as you would like them listed in the printed conference program. If there are changes, please write or fax them to us at your earliest convenience.

Please feel free to call us if you have any questions, concerns, or needs. This conference will address important issues, and we look forward to working with you. Again, contact us as needed at 800-937-4760 and ask for Cathie Parker.

See you in March!

Jan C. Simmons
Senior Program Development Specialist
Enclosures

Panhandle Mathematics and Science Conference
A Member of The Texas A & M University System
WTAMU Box 60208 Canyon, Texas 79016-0001 806-651-2626
Fax 806-651-2626

West Texas A & M UNIVERSITY
Division of Education

August 31, 2000

Paul Emekwulu
P.O.Box 2482
Norman, OK 73070

Dear Paul:

Math Paradise

Re: Panhandle Mathematics and Science Conference

Thank you for your proposal for the above upcoming conference on Saturday, September 30th, 2000 titled, *Math — Magic: Encouraging Mathematical Reasoning.*

I am delighted to inform you that your proposal has been accepted, and that we are in the process of constructing the schedule for the conference at this time. The schedule, with times and room numbers, will be posted to our web site at www.wtamu.edu as soon as we have it completed. There is an icon on the bottom left-hand corner of the webpage that will lead you to the Panhandle Math/Science Conference site. You will be presenting your session 1 time(s) and we suggest that you prepare for 30 participants in each session.

Registration for the conference will begin at 8 am in the Jack B. Kelley Student Center on the WTAMU campus where there will be a speaker packet waiting for you. Lunch is provided and we hope that you will enjoy your day with us. If for some reason you cannot be with us, please be kind enough to let me know as soon as you can so that arrangements can be made to cancel your session.

If I can be of any further help, please do not hesitate to contact me by email at cpurkiss@mail.wtamu.edu or by phone at 806-651-2618. I look forward to seeing you on the 30th of September.

Sincerely,
Chris Purkiss
Chairperson

There Could be a Book in You

Program Objectives

In this seminar, the participants will:

- Gain inspiration and motivation that could translate into action.
- Realize that dreams and intuition could be sources of book ideas.
- Realize that only three things can stop a dream book.
- Realize that affirmations could be used to activate our creativity.

Program Description

How many of you have ever thought of writing a book?
Are you working on a book now, or have you ever submitted a manuscript to a publisher?

Do you think you have special knowledge or skills that you would like to share with others?

Are you a good story-teller?

Can you tell stories in a manner that holds people's attention?

Are you a teacher and have unusual but creative ways of presenting ordinary classroom concepts?

Are you currently doing seminars based on your experiences, and you don't have a book covering your topics?

Do you write articles for newspapers or magazines? If yes, have you ever thought of building a book out of these articles?

Do you have ideas for a book but don't know how to put them

together?

Have you ever thought of collaborating with someone on a book project?

Do you have an idea of a particular cause that you would like to be remembered for?

If your answer to any of the above questions is *yes*, you need to write a book.

Emekwulu maintains that only three things can stop anyone from writing his or her dream book:

- lack of message,

- lack of faith in the message, and

- lack of faith in the messenger.

Past Engagements

- Norman Galaxy of Writers, Inc.
-
- Oklahoma City Writers, Inc.
-
- Canadian Valley Lions Club

- Metropolitan Library System, Mid-West City, OK

- Elk City Carnegie Friends of the Library, Elk City, OK

- Mid-Oklahoma Writers' Club

- Oklahoma Education Association

- Moore Association of Classroom Teachers, Moore, OK

About the Author

Originally from Nigeria, Paul Chika Emekwulu is an award-winning and international bestselling author. He is a co-author of an international bestselling book titled, *Unwavering Strength (Volume 2): Stories to Warm Your Heart & Soul.* When asked by the US Embassy officials in Lagos, Nigeria, why he petitioned for a non-immigrant entry visa into the United States, he said that he was coming to the United States to explore opportunity in publishing. Today, he is the author of six or more books including:

- *Mathematical Encounters for the Inquisitive Mind*
- *Getting to Know Triangular Numbers, Book One*
- *Getting to Know Fibonacci Numbers*
- *Divisibility Rules of Whole Numbers Made Simple*
- *Mathematical Explorations for Advanced Students*
- *Writing Down Your Dreams*: *Listen to Your Inner Voice and Change Your Life* etc.

He has made invited presentations for schools, and organizations including:

- Oklahoma State University, Stillwater, OK
- National Council of Teachers of Mathematics
- Oklahoma Education Association
- Moore Association of Classroom Teachers
- Kansas Association of Teachers of Mathematics
- Oklahoma Council of Teachers of Mathematics
- Oklahoma City Writers Inc.
- Norman Galaxy of Writers Inc.
- Black Liberated Arts Center, Oklahoma City, OK

- 7 Hawks Publishing
- Mid-Oklahoma Writers Club
- Newkirk High School, Newkirk, OK
- Washington High School, Washington, OK
- Booker T. Washington High School, Tulsa, OK
- Oklahoma City Community College (*Upward Bound*)
- Michael Price School of Business of The University of Oklahoma

He is a member of Oklahoma Council of Teachers of Mathematics and Central Texas Council of Teachers of Mathematics.

He independently developed a mathematical formula connecting triangular numbers and numbers of the Fibonacci sequence. He has been a guest on several radio stations across the United States.

Useful Links

Meet the Co-Authors of *Unwavering Strength* Here
http://unwaveringstrength.com/co-authors/#PaulE

Social Media

Connect with me on Linkedin
https://www.linkedin.com/profile/view?id=65353103&trk=nav_r
esponsive_tab_profile

Follow me on Twitter
https://twitter.com/pemekwulu

Befriend me on Facebook
https://www.facebook.com/authorpaulchika.emekwulu

Media Interviews

Listen to Paul's Interview with Cathryn Taylor of *Edge Magazine*
http://www.blogtalkradio.com/edgemagazine/2014/03/12/edge-
inner-views-with-cathryn-taylor-and-paul-emekwulu

Listen to Paul's Interview with Anya Sophia Mann
http://anyasophiamann.com/quantum-alchemy/unwavering-
strength/paul-emekwulu.php

www.ingramcontent.com/pod-product-compliance
Lightning Source LLC
Chambersburg PA
CBHW051324170526
45166CB00002B/681